DAMP PROOF COURSE DETAILING

Second edition

John Duell and Fred Lawson

The Architectural Press Ltd
London

Damp Proof Course Detailing

Second edition

DAMP PROOF COURSE DETAILING

Second edition

John Duell and
Fred Lawson

The Architectural Press Ltd
London

Acknowledgements

Much of the reseach for sections 1-8 of this book was carried out while preparing a paper presented at a conference of the Building Materials Section of the British Ceramic Society held in London in November, 1974. The book is partly based upon a survey of architects, engineers, manufacturers, etc, and Mr Duell's thanks are due to all the individuals and institutions who contributed information. For many ideas included here, the author would particularly note his indebtedness to Donald Foster RIBA, and his excellent booklets published by Structural Clay Products Ltd, and to Cecil Handisyde RIBA, and his team responsible for 'Everyday Details' published in the Architects' Journal.

Thanks are also due to the Agrément Board for their advice on the section on *Dampness in Buildings* and to Davis, Belfield and Everest, Chartered Surveyors, for revising the data on costs of remedial work, in particular to Mr S. Johnson who again revised the costs for this second edition.

Note on second edition

Several important developments have occurred since the first edition of this book was published.

The trend towards more thermally efficient buildings has caused an increase in the use of various forms of cavity insulation. The filling of cavities contradicts the philosophy of cavity wall detailing. Except in the most sheltered sites, any decision to use cavity insulation must be very carefully considered and detailed.

The design of brick clad concrete framed buildings has become more sophisticated—perhaps too sophisticated—in order to overcome problems of shrinkage, expansion and thermal movement. Stainless steel angles and other devices have been introduced to support the brick outer skin while allowing for differential movement. As many of these devices normally bridge the cavity, they must be flashed with a dpc; which can be difficult.

More preformed cloaks, in a range of materials, are now available to simplify corners, roof to wall junctions and the like. Although these cloaks help provide a better job, careful workmanship during installation is still essential—and difficult to get.

First published in book form in 1977 by
The Architectural Press Ltd: London

Second edition 1983

ISBN 085139 149 4 (paper)
ISBN 085139 150 8 (cloth)

Typeset by Phoenix Photosetting, Chatham

Printed in Great Britain by Diemer & Reynolds Ltd, Bedford

Contents

1 Introduction

Correct selection, detailing and installation

Many current failures in building construction can be traced to faulty damp proof courses—faulty choice of material; faulty detailing; or faulty installation. JOHN DUELL, an architect in private practice in Colchester, has carried out a study of dpc installation, the reasons for failure, and appropriate solutions, and suggests that there are three main reasons for failure:

- **First, existing textbooks and references concentrate on** *rising* **damp, and give insufficient attention to the much more challenging problem of** *downward rain penetration* **in medium to high-rise exposed buildings.**
- **Second, architects tend to draw a few cross-sections through relatively simple parts of the dpc installation, and neither think out what happens at the awkward junctions, nor adequately communicate clear instructions to the builder (for instance, three-dimensional formation and installation drawings).**
- **Third, the installation of dpcs on site is often carried out in a haphazard fashion with little direction from site supervisory staff.**

This short book combines two sets of articles, previously published in The Architects' Journal in 1976 and earlier, which have been brought together in a revised and updated form so as to provide a truly comprehensive coverage of the selection, detailing, specification and installation of all dpcs in a single publication. As well as dealing with dpcs themselves, they also discuss, in general terms, dampness in existing buildings which might be related to or confused with dampness due to defective dpcs. Notes on this aspect of the problem have been prepared by FRED LAWSON, formerly a lecturer in environmental studies at the University of Surrey, and the author of several Architectural Press books.

1.1 Reasons for failure

As in many other aspects of construction, there is a need today to re-examine the selection, detailing, specification and installation of damp-proof courses. A considerable number of building failures relating to dpc work have been noted, particularly by the BRE Advisory Service (AJ 5.2.75 pp303-308; CI/SfB (9–) (S)*. The failures have been attributed to both inadequate detailing, bad workmanship, and defective materials. Some people blame architectural education for ignoring building construction while others blame lower standards of workmanship on site. In fact, the reasons for dpc failures are many and varied. The higher buildings erected in the past two decades have meant that many dpcs are now actually required to perform their designed function (many dpcs in low-rise traditional buildings are never actually subjected to really damp conditions). All new types of construction have been the subject of water penetration in connection with dpc work, from load-bearing brickwork/blockwork to industrialised concrete-panel buildings.

1.2 Existing guides

Most importantly, relevant Standards on damp-proof course materials and construction are in many cases out-of-date, out-of-touch and too generalised to be sufficiently useful. BS 743:1970: *Materials for damp-proof courses*[5] does not include some of the most commonly used types, nor does it give details of the physical properties of dpc materials which are critical if an intelligent choice is to be made. And CP 121:Part 1:1973: *Code of practice for wallings*[4], like its predecessor in 1951, is perhaps over-concerned with traditional dpc details designed to withstand rising damp, and gives insufficient attention to damp-proofing details to withstand penetration by rainwater. Some of the suggested details in the Code of Practice could, in the author's opinion, lead to damp penetration while some of the recommendations on standards of workmanship are unrealistic and seldom obtainable on site. Unfortunately, construction textbooks and other publications are equally inadequate with regard to damp-proof courses.

1.3 This book

It is hoped in this brief book to outline an approach to dpc selection, detailing, specification and installation that will lead to more trouble-free buildings. It will concentrate on those aspects which are inadequately covered at present, and suggest an approach as much as recommend actual details. Suggestions are made on specific dpc details but it is hoped that this information will make the reader really think about his particular dpc problem.

The series has been written to be of assistance to all the members of the building team—architect, structural engineer, quantity surveyor, clerk of works and contractor—as responsibility for dpc work is shared among all of them; and it follows a workstage approach:

- general considerations
- selection of most suitable materials
- detailing of dpc in chosen materials
- specification of dpcs
- site installation and supervision

Finally, there is a comprehensive list of available flexible dpc materials.

* See BRE Digest 176, April 1975, HMSO.

2 General design considerations

2.1 Function of a dpc

Dpcs are necessary at many locations in a building to protect critical parts of the structure and the internal environment. There are three basic ways in which damp-proof courses are used:

- to resist damp penetration from below
- to discharge water from above
- to resist horizontal water penetration (eg at window jambs).

These three different functions of dpcs, **1**, call for quite different approaches to design and construction. Lack of understanding of this basic functional difference has been the root cause of many dpc problems.

2.2 Taking into account exposure

With the second and third of the functions listed above (resistance to downward and horizontal rain penetration) the degree of exposure to which the building face is subjected is a critical factor in dpc design.

There may be two fundamental approaches to this factor. One is to argue that there ought to be a single standard of cavity wall construction (and associated dpc design) which will work anywhere in the UK, and which designers can apply with confidence and without too much complex analysis. This is usually the official BSI and BRE approach to cavity wall design. The alternative is to adapt dpc design to exposure conditions; and this is the approach taken in the present series.

BRE driving-rain index

The publication *Driving-rain index*, **2**, by R. E. Lacy[21] (HMSO 1976; £2·60) superseded the previously published BRE Digest 127 (HMSO 1971; 6p); and while it is intended primarily to estimate exposure for the purposes of cavity-fill insulation, the indexes given could also usefully be applied to dpc design. The procedure for estimating the exposure category for any particular building is this:

1 Locate the building site on the map, and read off the annual mean driving rain index from the contours, **2**.

2 Decide whether the surrounding topography is generally level, more than ordinarily exposed, or more than ordinarily sheltered.

3 Decide which of the four categories of terrain roughness given in the publication is the appropriate one for the site (ranging from 'open country with no wind breaks' at one extreme, to 'heavily protected by nearby trees or buildings' at the other).

4 Now any face of the building, at any particular height above ground level (up to 10 metres), can be given a driving-rain index of 'sheltered', 'moderate', or 'severe' by referring to the set of tables A to H in the guide[21].

BS 5618:1978: *Code of practice for thermal insulation of cavity walls by filling with UF foam* has further developed the calculation of exposure and defines an exposure index based on three factors: geographic, topographic, and terrain roughness. For general definition of the degree of exposure it is not, in the author's opinion, as useful as the original method defined by Lacey. The procedure adopted was previously outlined by the Agrément Board in their *Information sheet No 10* (Agrément Board 1977).

Rough estimation

For quick preliminary estimation of the situation, the following alternative simple guide may help.

1 Locate the building site on map **3** (from BRE Digest 127), but consider all areas within 8 km of a coast or estuary as one category of exposure higher than that shown on map.

2 Consider the site topography as indicated in **4**; and upgrade the exposure category decided in step 1 in the case of 'severe' site-form conditions (eg 'sheltered' would become 'moderate' if building were on hilltop as in **4a**).

3 Consider building height as indicated in **5**, and adjust category of exposure if necessary. Low-rise building will benefit from shelter by hills, trees, walls and other buildings, and the actual exposure rating may therefore be one degree *lower* than that indicated by the site exposure given under step 1. On the other hand, if a building rises significantly above the prevailing height of buildings and other obstructions around it, then the actual exposure rating will probably be a degree *higher.*

2.3 Taking into account construction

For any given building face, we now have an exposure rating of sheltered, moderate, or severe.

But another critical factor in the design of dpcs for a particular building is related to its basic type of construction (see **table I**). Taking extremes, for illustrative purposes:

- A fairfaced blockwork wall, **6a**, will have a much higher risk of water penetration, and consequent higher demand on dpcs, than a rendered brick cavity wall, **6b** (in areas of severe exposure some cavity walls in concrete blockwork have been found to be inadequate as rain blows through the outer skin, across the cavity and on to the inner skin).
- Impervious panel types of construction, **7a**, often direct large volumes of water towards joints, where a high standard of dpc detailing is therefore required; conversely, an absorbent brick wall, **7b**, acts as a collector of rainwater, holding it within the fabric and releasing it gradually, so that the joints and flashings are therefore not usually so critical.

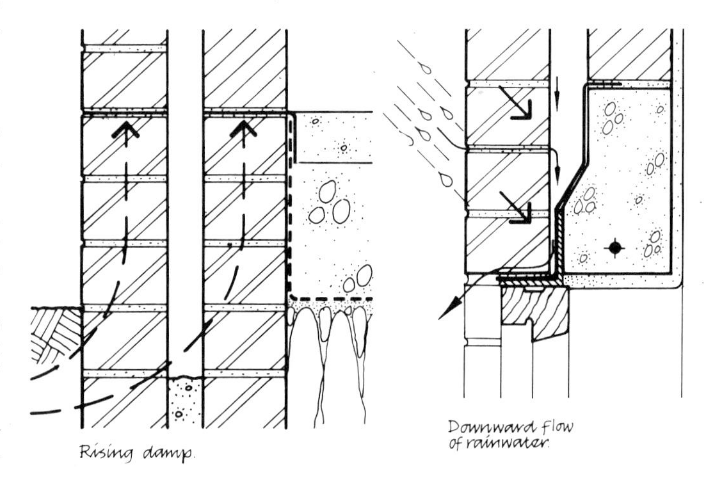

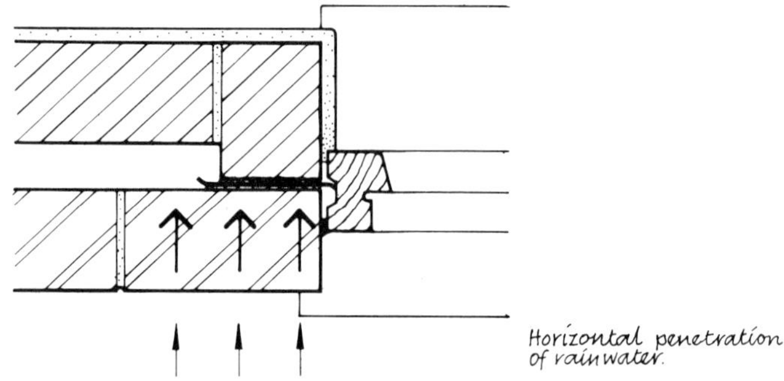

1 *Three different functions of dpcs, which must be taken into account. Each raises its own problems of selection and detailing.*

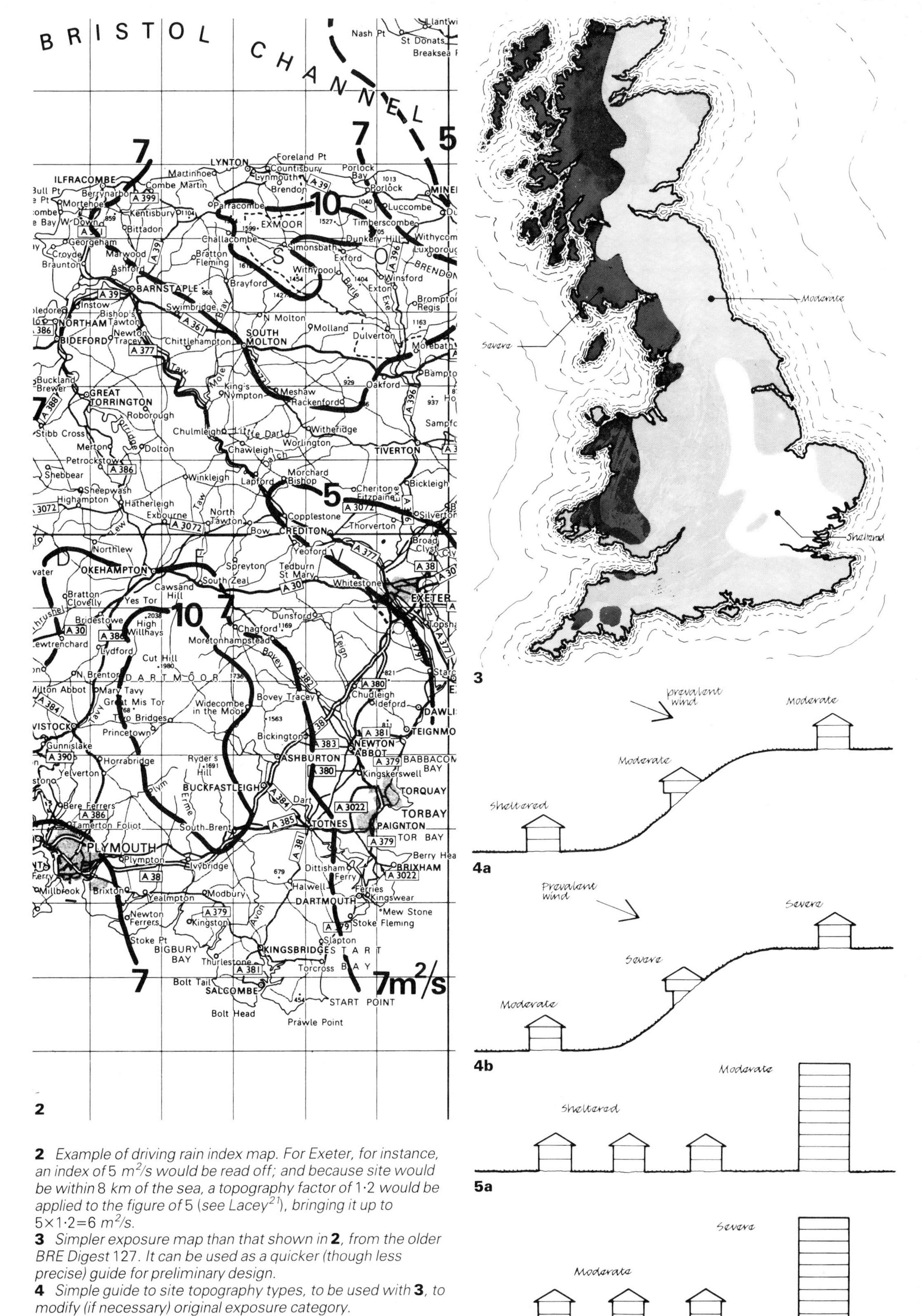

2 *Example of driving rain index map. For Exeter, for instance, an index of* 5 *m^2/s would be read off; and because site would be within* 8 *km of the sea, a topography factor of* 1·2 *would be applied to the figure of* 5 *(see Lacey*[21]*), bringing it up to* 5×1·2=6 *m^2/s.*

3 *Simpler exposure map than that shown in* **2**, *from the older BRE Digest* 127. *It can be used as a quicker (though less precise) guide for preliminary design.*

4 *Simple guide to site topography types, to be used with* **3**, *to modify (if necessary) original exposure category.*

5 *Simple classification of building heights, which can be used to either upgrade or reduce original exposure category from* **3**.

Table I Typical examples of higher and lower risk constructions*

Higher risk
Solid masonry walls (especially if thin)
Fairfaced block cavity walls
Panel systems (eg precast concrete)
Brick clad timber frame
Concrete frame/brick infill (if frame is exposed)

Lower risk
Brick cavity walls
Rendered block cavity walls
Timber and tile clad timber frame
Concrete frame/brick infill (if frame behind cavity)

* These are generalised categories only, and each particular case needs more detailed assessment on its merits—eg, if there is a possibility of rendering cracking, a rendered 'low risk' wall could become 'high risk'.

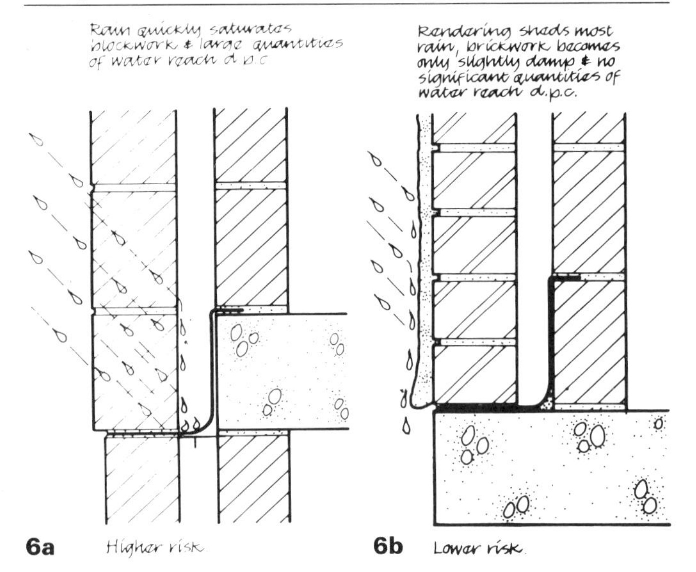

6a Higher risk **6b** Lower risk

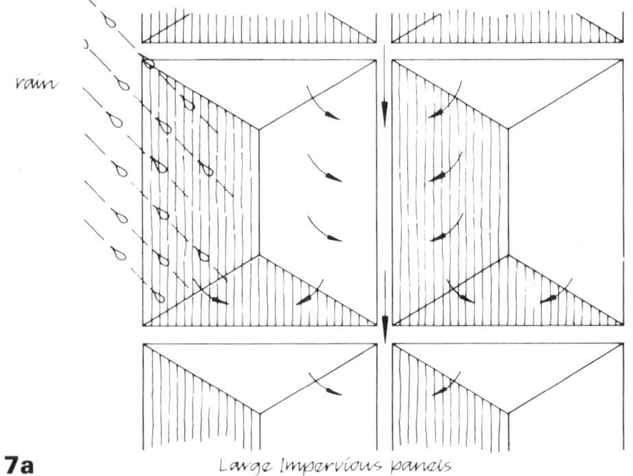

7a Large impervious panels

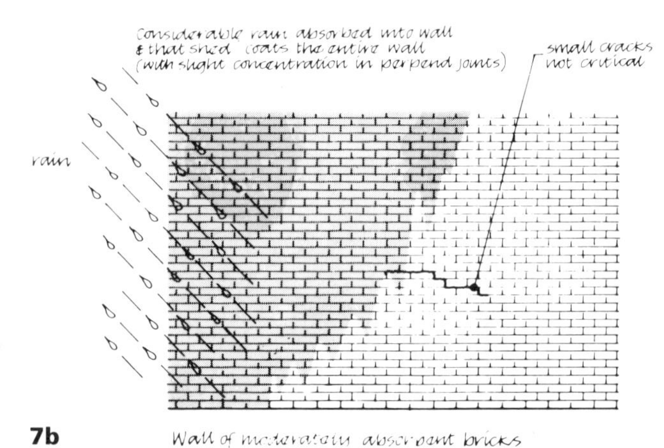

7b Wall of moderately absorbent bricks

6, 7 *Regardless of which method has been used for estimating exposure, the next step is to assess effect of construction on dpc performance.* **6a** *and* **7a**, *for instance, are likely to place greater demand on dpcs and flashings, than* **6b** *and* **7b.**

• Considerable discussion has taken place on the relative rain penetration experienced by open-pored bricks versus dense bricks. While general conclusions from laboratory tests can be misleading, the British Ceramic Research Association has shown that dense bricks well-laid in high strength mortar are more resistant to water penetration. These tests, however, showed that the most important factor in determining the relative rain penetration was the quality of workmanship in laying the bricks, with partially filled perpend joints the main culprit. (See BCRA Special Publication 60, 1968, *The performance of walls built of wirecut bricks with and without perforations*).[24] Workmanship is equally important in fairfaced blockwork and more difficult to achieve.

2.4 Decision on basic standard of dpc

Given the above factors (degree of exposure of site and building; type of construction) the designer is now in a position to decide upon the basic standard of dpc detailing and specification. In **table II** these factors are correlated and recommendations made. Generally the designer will only be in a quandary about the standard of detailing necessary in the middle range (ie code B) and this will require the use of judgment. In matters regarding the weather-proofing of buildings, it is wise to err on the side of caution. To illustrate the implications of the three different exposure conditions for dpc performance, **10** gives an indication of the amount of rainwater penetrating into a typical brick cavity wall after being subjected to wind-blown rain for a considerable period.

Table II Suggested correlation of factors to decide basic standard of dpc detailing

Site category (from para 2.2)	Construction category (from para 2.3 and table I)	Building height and suggested dpc standard A, B or C*		
		Low rise (1-3 storeys)	*Medium rise (4-7 storeys)*	*High rise (8 storeys and over)*
'Sheltered'	lower risk	A	A	B
exposure site	higher risk	B	B	C
'Moderate'	lower risk	B	B	B
exposure site	higher risk	B	C	C
'Severe'	lower risk	B	C	C
exposure site	higher risk	C	D	D

* Basic dpc standards referred to above:
A Dpcs specified with no elaboration of detailing necessary; and only basic, normal standard site of installation.
B Dpcs carefully specified with some elaboration of detailing necessary at vulnerable positions; and with above-normal, good standard of site installation.
C Dpcs very carefully specified in regard to each situation; with thorough consideration of detailing, and high standard of site installation.
D In these cases consideration should be given to a change in basic type of construction from higher risk to lower risk (see table I), and if this is not possible, then an extra high dpc standard must be detailed, specified and installed.

2.5 Cavity fill insulation and dpc detailing

The practice of filling cavities with various types of insulating material is spreading and Amendment 2 of the Building Regulations will increase this trend. Until recently British Standards and Agrément certificates limited the use of cavity insulation to fairly sheltered sites, but some types such as mineral fibre are now recommended for high rise buildings in quite exposed situations (exposure index of up to 120). In assessing suitability, careful reference to exposure in BS 5618 and manufacturers' literature may be adequate in most cases, but in the author's opinion the designer must consider very carefully the use of cavity insulation.

Particular attention to detailing (and to installation on site) is required around openings, ring beams, parapets, and roof abutments. These are not well covered, either in Agrément certificates or in manufacturers' literature (see later sections on detailing and installation).

If **table II** above is consulted and the particular building is assessed as if the cavity were *not* filled, then the following conclusions will guide the designer to a cautious approach to the specification of cavity fill insulation:

Standard A Cavity insulation should cause no problems provided a reasonable standard of workmanship is maintained on site.

Standard B Cavity insulation may be satisfactory given a good standard of workmanship and particular attention at detailing stage to dpc details and cavity insulation round openings.

The author would not himself install cavity insulation above five storeys in any case.

Standard C 50 mm cavity must be maintained,

For some types of cavity insulation, such as polystyrene slabs, it is recommended that a cavity is maintained, and if a 50 mm cavity is provided there are no exposure restrictions. Care should still be taken around window openings and in other similar positions.

2.6 Location of dpc in building structure

The different parts of any building suffer different degrees of exposure to rainwater and damp, **8**, **9a**. It is important to realise this so that after a decision has been made regarding the basic standard of dpc detailing, the structure is then considered for areas that are particularly vulnerable, such as parapets, flush eaves, and retaining walls; junctions between different parts of the buildings; and all cavity bridges (eg ring beams). Wide overhangs provide considerable protection and can greatly reduce exposure, but are unfortunately seldom designed in the UK. Examples of suspect detailing and selection of bricks leading to serious efflorescence are shown in **9b** and **9c**.

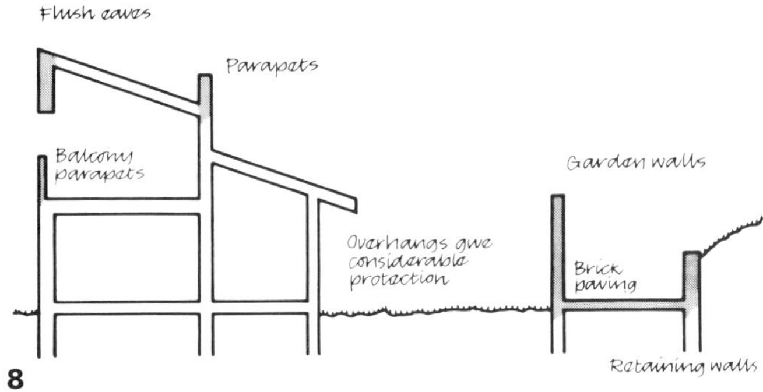

8

8, 9a *Particularly vulnerable areas in building structure, where damp penetration requires special attention include flush eaves, balcony parapets, roof parapets, garden and retaining walls, pavings (in the case of brick paving, of course, dpcs are not relevant; the precautions would be in terms of quality of paving).* **9b** *Serious efflorescence has here occurred at a secret gutter to a mansard roof, at a parapet to the flat roof and to a free-standing wall.* **9c** *Even in traditional construction, careful consideration must be given to junctions of different materials and elements if efflorescence is to be avoided.* **10** *Damp penetration into cavity after considerable period of wind-blown rain, illustrating effect of different exposure conditions.*

9a

9b

9c

Slightly damp on inner face, mostly at perp. joints.

Dp tray virtually dry

(a) SHELTERED

Most of inner skin damp, but water not running down

Dp tray damp

(b) MODERATE

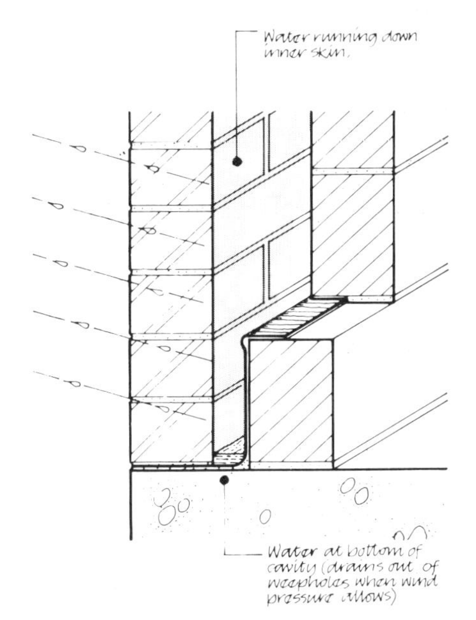

10

3 Selection of suitable dpc material

3.1 Present information inadequate

BS 743:1970: *Materials for damp-proof courses*[5] gives the 'official' list of dpc materials available in the UK; but this has many omissions, and other sources have attempted to give a more complete summary but with no organised attempt to define their properties and consequently their performance. This lack of information has hitherto made the intelligent selection of a suitable dpc material almost impossible.

3.2 Classification of dpc materials

Table III includes all materials in general use for dpcs, classified into five basic groups:
rigid materials
malleable metals
bitumen-based materials
polymer-based materials
in situ coatings.
The rigid materials are only suitable for rising damp and a very limited number of other damp-proof course applications; and the malleable metals are very expensive (though they remain the safest method for difficult spots). In situ coatings call for particularly diligent workmanship. Consequently, functional and economic considerations have meant that bitumen and polymer-based sheet materials account for most present-day dpc installations. These two types of dpc are often grouped together under the term 'flexible damp-proof courses'.
Note: properties vary considerably between manufacturers and final selection should be made from actual samples.

3.3 Properties of dpc materials

Traditional dpc materials have not been studied with regard to their physical properties although the composition is specified in some detail in BS 743[5]. The BRS in the early 1960s carried out tests on a number of new dpc materials but they were not very comprehensive. Since 1967 several dpc products have been certified by the Agrément Board who do extensive tests on relevant physical properties (eg tensile strength and elongation at break; tear strength; high and low temperature flexibility; and resistance to compression). No attempt has been made to relate these tests between the various certified products, and they are not, therefore, very useful to the designer.

Tests for durability are a difficult matter and careful analysis on the new materials is essential. Of equal importance, however, are properties relating to workability (ie ability to be made into shape in cold and warm weather, ability to keep that shape without restraint, forming of laps, and ease in forming junctions and corners). A test programme is needed on the properties of dpc materials that will give quantifiable results, so that comparisons can be made between materials and between manufacturers. These test would call for close co-operation between scientist, designer and user (ie the contractor and his employees) to ensure that the tests are really relevant.

In the absence of these, the author has attempted in **table III** to make comparisons. It must be emphasised that these are a 'first stab' at a performance chart that can be used by designers in selecting a suitable dpc material for a particular project and a particular location within that project, and that the information is provisional—feedback from manufacturers, designers, builders and research or testing organisations would be valuable. (This plea was written for the first edition and, alas, has gone unheeded).

3.4 Suggested applications of dpc materials

In selecting a dpc material the designer must study the properties of various dpc materials in relation to the particular situation. There is often confusion in discussions regarding the various dpc situations and **11** defines the terminology used in this series.

Dpcs for rising damp
Here the selection is fairly simple as workability is not usually a significant factor except when a dpc has to be stepped and/or jointed to a damp-proof membrane; and performance requirements are straightforward. Bitumen-based materials have been used successfully for years although they can extrude bitumen and be unsightly, **11a** (in high-load situations they should be

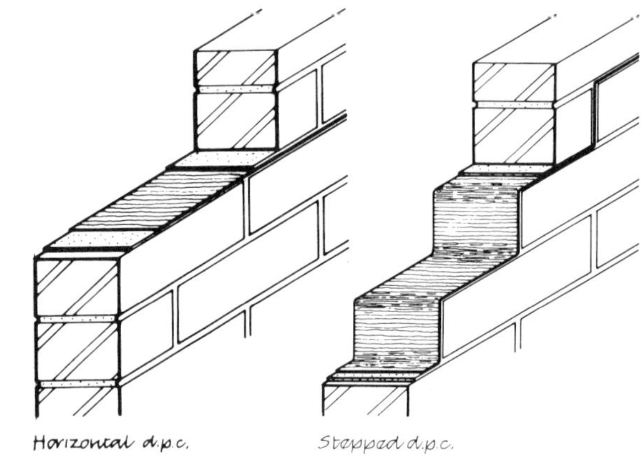

11 *Definition of terms which will be used throughout this study.*

11a *An example of unsightly bitumen extruding from a bitumen-based horizontal dpc. Note also apparent movement at dpc.*

Table III Available dpc materials and their relative properties

Group	Material (with British Standard relevant to material as such)	Relevant British Standard for use as dpc	Minimum thickness (mm)	Minimum weight (kg/m²)	Resistance to compressive loading	Resistance to lateral loading (tests on this property are difficult to interpret)	Freedom from extrusion	Durability		Ease of handling on site, and ease of storage	Workability		Ability to be made into shapes		Resistance to damage (at 20°C: may differ in relative terms at other temperatures)				
								Covered	Exposed		High temperature	Low temperature	Ability to keep shape without restraint	Ease in forming corners and junctions	Extensibility	Tearing (resistance to 'first tear', not continuation: and taken in weakest direction of material)	Puncture	Preforming readily available	Ease of forming and sealing laps
Rigid materials	*Brick* (to BS 3921)	BS 743	Two courses; 150 mm	—	•••	•••	•••	•••[1]	•••	•	Only suitable for rising damp function, in carefully selected situations								
Rigid materials	*Slate* (to BS 3798)	BS 743	Two courses; slate 4 mm thick	—	••	••	•••	•••[1]	•••	•	See note above								
Malleable metals	*Lead* (to BS 1178) Lead must be coated with bitumen	BS 743	1·80 (code no 4)	19·50	••	•••	•••	•••	•••	•	•••	•••	•••	••	•	•••	•••	No	•
Malleable metals	*Copper* to BS 743	BS 743	0·25 (if exposed, must be thicker see BS 743)	2·28	•••	••	•••	•••	•••	•	•••	•••	•••	••	•	•••	•••	No	•
Malleable metals	*Zinc* (to BS 849) Zinc may require coating with bitumen	None (but see CP 143 part 5)	0·81 (14 g)	5·40	•••	••	•••	•••	••	•	••	••	•••	•	•	•••	•••	Yes	•
Malleable metals	*Aluminium* (to BS 1470)	None (but see CP 143 part 15)	0·91 (20 g)	2·80	•••	••	•••	•••	••	•	••	••	•••	•	•	•••	•••	Yes	•
Malleable metals	*Zinc/lead* (proprietary product)	None (but see zinc, above)	0·60	4·30	•••	••	•••	•••	••	•	••	••	•••	•	•	•••	•••	Yes	•
Malleable metals	*Pressed steel*[5]	None	22-24 g	—	•••	•	•••	••[6]	••[6]	•	*only preformed*		•••	••	Nil	•••	•••	Yes	••
Bitumen-based products	*Bitumen/hessian base**	BS 743 type A	2·5[4]	3·80	•	••	•	••	•	•	••	•	•	••	•	••	•	No	•••
Bitumen-based products	*Bitumen/fibre base**	BS 743 type B	2·5[4]	3·30	•	••	•	••	•	•	••	•	•	•	•	•	•	No	•
Bitumen-based products	*Bitumen/asbestos base*	BS 743 type C	2·5[4]	3·80	•	••	•	••	•	•	••	•	•	•	•	•	•	No	••
Bitumen-based products	*Bitumen/hesian base/lead**	BS 743 type D	3·0[4]	4·40	•	••	•	•••	••	•	••	•	••	••	•	••	••	No	•••
Bitumen-based products	*Bitumen/fibre base/lead**	BS 743 type E	3·0[4]	4·40	•	••	•	•••	••	•	••	•	••	•	•	•	•	No	••
Bitumen-based products	*Bitumen/asbestos base/lead*	BS 743 type F	3·0[4]	4·90	•	••	•	•••	••	•	••	•	••	•	•	•	•	No	••
Bitumen-based products	*Bitumen/hessian base/aluminium*	None (but superior to type A above)	3·0[4]	4·10	•	••	•	•••	••	•	••	•	••	••	•	•	••	No	•••
Bitumen-based products	*Bitumen/fibre base/aluminium*	None (but superior to type B above)	3·0[4]	4·90	•	••	•	•••	••	•	••	•	••	•	•	•	•	No	••
Polymer-based[2]	*Polythene**	BS 743	0·46	0·48	•••	••	•••	••	•	•••	••	••	•	•	•••	••	•	No	•
Polymer-based[2]	*Polypropylene* (proprietary product)	None	0·8 and 1·5	—	•••	••	•••	••	•	•	••	••	•	••	••	••	••	Yes	•
Polymer-based[2]	*Pitch polymer**	None	1·27	1·50	•••	•••	•••	••	•	•••	•••	•••	•	•••	••	••	••	Yes	•••
Polymer-based[2]	*Bitumen polymer* (proprietary product)	None	1·25	1·60	•••	•••	•••	••	••	•••	•••	•••	•	•••	••	••	•	No	•••
Insitu coatings	*Mastic asphalt* (to BS 1097 & BS 1418)	BS 743	12·0	—	•	••	•	••	•	•	•••	•••	•••	••	Virtually nil	N/a	•[3]	N/a	N/a
Insitu coatings	*Epoxy resin/sand*	None	6·0	—	••	•••	•••	•••	•	••	•••	•••	•••	•••	As above	N/a	•[3]	N/a	N/a
Insitu coatings	*Bitumen/rubber/glass fibre*	None	1·5	—	•	•	•	••	•	••	•••	•••	•••	•••	•	N/a	••[3]	N/a	N/a
Insitu coatings	*Pitch epoxy/glass fibre*	None	1·5	—	••	•••	•••	•••	••	••	•••	•••	•••	•••	•	N/a	••[3]	N/a	N/a

Key to symbols: • below average; •• average; ••• above average; N/a not applicable. Asterisk (thus: Polythene*) denotes types most frequently used.

Notes General: properties of materials may differ according to individual manufacturer. The alternatives should be carefully examined physically: some manufacturers produce materials to higher specifications.

1 Inherently durable in themselves but susceptible to fracture under building movement.

2 A number of proprietary products in this group have Agrément certificates: a list is published in the Product Guide (Section 8).

3 Resistance to puncture is not applicable, so resistance to fracture is assessed.

4 Approximate thickness: not part of specification.

5 Pressed steel is fairly rigid; malleable skirt or apron of lead is often attached. Steel is available galvanised or stainless.

6 Stainless versions would be above average.

used with extreme care). Their relative thickness makes them difficult to conceal in the mortar joint, particularly at laps. For loadbearing brickwork, brick and slate are possible, but they may not be completely reliable. The polymer group stands out for its ability to resist compression, its relative thinness and freedom from extrusion. In lightly loaded walls (eg garden walls), it is very important that a dpc material which helps give the wall high resistance to lateral loading is selected. Brick or perhaps slate will give adequate resistance of this kind; but generally such walls will have to be carefully designed to be stable from dpc upwards.* Lateral movement at dpc level occasionally occurs on normally loaded walls too, and this is sometimes blamed upon the dpc material. This is doubtful, but again a material able to develop good tensile bond with the brickwork, and with good resistance to tearing or fracture, may help to minimise the problem.

Dpcs for rainwater (downward or horizontal penetration)

These are the second and third functions mentioned in section **2.1**; and here the selection will be more difficult than in the case of rising damp, as many factors will be involved.

The *durability* of the material must be considered (as, of course, it should be with rising damp as well). The malleable metals are outstanding in this regard, but the lead-cored bitumen-based materials are also well-proven. If the dpc is partially exposed, the suitability of the material must be particularly carefully analysed. Some of the pitch and bitumen polymer manufacturers claim good durability in exposed situations, but as they are relatively new materials their actual performance is difficult to assess. The formation and sealing of laps is very important for dpcs discharging rainwater. Bitumen-based materials are fairly easy to seal (although heating is necessary), while pitch and bitumen polymers can be effectively lapped and sealed without heating by using adhesives. The laps in polythene dpcs are quite difficult to seal effectively against downward penetration of rain water.

The various *workability* factors, however, are of paramount importance for such applications as cavity dp trays. Malleable metals are workable but lead is very heavy and expensive; and fabrication is a time-consuming process with all metals, much of it best done off-site or in a site hut. The bitumen-based materials are workable but are still quite heavy and require considerable care in fabrication, particularly at low temperatures. In the polymer group, characteristics differ to some extent as polythene is light and easy to handle but it is very difficult to place adequately because its resilience will not allow it to keep a shape without restraint (this is true to some degree of several bitumen-based types too). Pitch polymers are easier to place in position as they are less resilient. In situ coatings are by their nature very workable but laying is a time-consuming process and effectiveness depends on consistently good workmanship.

Resistance to damage

It is vitally important that a dpc material withstands damage during construction. Low temperature flexibility, tensile strength and tear strength are obviously important as is resistance to puncture. The main danger to the dpc is the essential—though often neglected—process of cleaning out of cavities of mortar droppings (see section **6.4** for details). Malleable metals have the highest resistance to damage but with lead and aluminium cored bitumen dpcs the thinness of the core metal makes them more susceptible to damage, with the hessian laminated materials being the toughest. Polymers are fairly tough, especially in relation to their weight and particularly at lower temperatures. They can be damaged, however, and reasonable care during installation is essential. Once a tear or cut is made in polymer materials, continuation is much easier. The in situ coatings that can be reinforced with glass fibre are very resistant to damage, but asphalt and resin/sand can be damaged quite easily.

photo by courtesy of Anderson roofing

12 *Example of widely used flexible dpc (black polythene). Mortar bed omitted for clarity.*

Table IV Economics of dpc materials and installation

Material		Basic cost of material[3] (polythene taken as 100)	Installed cost of two dpc types (polythene taken as 100[1])		Total cost of dpc work for 3 blocks of 6 storey maisonettes in loadbearing brickwork: 36 dwellings costing approximately £1,100,000[4]	
			Horizontal dpcs	Cavity trays 333 mm girth	Total (polythene taken as 100)	Percentage of contract sum
Malleable metals	*Lead* (1·80 mm)[5]	3,109	N/a[2]	1,063	929	1·10
	Zinc/lead (proprietary product)	1,905	N/a[2]	744	658	0·80
Bitumen based	*Bitumen/hessian*	246	157	143	145	0·18
	Bitumen/hessian/lead	917	400	324	331	0·40
Polymer based	*Polythene*	100	100	100	100	0·12
	Pitch polymer[6]	373	198	172	177	0·35

Notes

1 The index of prices is based on rates ruling at September 1982.
2 N/a means this material is not applicable, usually, to horizontal dpc situations.
3 Material costs vary widely depending on discounts available.
4 Includes horizontal and vertical dpcs and cavity trays. Pitch polymer is assumed to have been used for horizontal dpcs, instead of the material listed, in cases marked N/a (in these cases the listed material is not suitable for horizontal dpcs—see note 2).
5 Installed costs for lead include painting both sides with bitumen.
6 Preformed angle and cloaks are available; the installed costs of these range from £6·20 to £8·50. They are quite expensive but so is the true cost of forming on site.

* See *Hard Landscape in Brick*, chapters 2 and 10. Published in book form by The Architectural Press 1977.

Table V Suggested applications of dpc materials listed in table III

	Materials (see table III)	Horizontal dpc at base of wall, normally loaded	Horizontal dpc at base of wall, lightly loaded	Horizontal dpc at base of garden wall	Horizontal dpc under coping	Stepped dpc at end of wall, etc	Horizontal dpc under sill	Cavity dp tray over lintel	Cavity dp tray to drain cavity wall etc (simple situations)	Cavity dp tray to drain cavity wall etc (complex situations)	Cavity dp tray to drain parapet cavities	Dp tray to chimney stacks	Vertical dpc at window jambs, etc	Abutment of pitched roof with cavity wall[2]
Rigid	*Brick*			●●										
	Slate			●●										
Malleable materials	*Lead*				●							●●		
	Copper				●							●●		
	Zinc				●●							●●		
	Aluminium													
	Zinc/lead				●●							●●		
	Galvanised (or stainless) pressed steel/lead							●	●	●				●●
Bitumen based products	*Bitumen/hessian*	●			●	●	●	●	●	●			●	
	Bitumen/fibre	●			●								●	
	Bitumen/asbestos	●	●●[1]	●●[1]	●								●	
	Bitumen/hessian/lead	●			●	●	●	●	●●	●●	●		●	
	Bitumen/fibre/lead	●	●	●	●								●	
	Bitumen/asbestos/lead	●			●								●	
	Bitumen/hessian/aluminium	●			●	●	●	●	●	●	●		●	
	Bitumen/fibre/aluminium	●			●								●	
Polymer based	*Polythene/polyethylene*	●●				●●	●●	●●	●				●●	
	Polypropylene						●	●●	●●	●●				●●
	Pitch polymer	●●	●	●	●	●●	●●	●●	●●	●●	●●		●●	●●
	Bitumen polymer	●●	●	●	●	●●	●●	●●	●●	●●	●●		●●	
In situ coatings	*Mastic asphalt*				●									
	Epoxy resin/sand				●									
	Bitumen/rubber/glass fibre									●				
	Pitch expoxy/glass fibre (sanded)		●							●				

Key to symbols: ● suitable; ●● especially suitable for functional or installational advantages.

Notes
Properties of materials may differ according to individual manufacturer. The alternatives should be carefully examined physically: some manufacturers produce materials to higher specifications.
1 One bitumen/asbestos dpc has been specifically developed for these locations and carries an Agrément certificate: see Product Guide Section 8).
2 Preformed cloaks available in materials noted for this situation.

3.5 Economic considerations

In an ideal world building materials and components, especially dpcs, would be chosen purely on a functional basis; but usually budgets are limited. **Table IV** gives costs of some of the most commonly used dpc materials.

As with so many savings of this type, selecting a cheaper dpc may in some cases be a tragically (and expensively) short-sighted decision.

3.6 The right material for the job

It is hoped that the above discussion will help the designer choose a suitable dpc for a particular project and location within that project. **Table V** suggests dpc material applications in relation to typical locations with particularly suitable materials highlighted. Like all selection, the final choices will probably mean compromise. For example, there is the understandable desire to have only one dpc material for a given project to avoid confusion on site. This can often be done, but care must be taken to ensure that the selected dpc is suitable for the various situations.

It is the practice of some firms to have a standard office specification for dpcs. While this is again understandable, the above discussion should serve as a warning that each job must be carefully reconsidered with regard to the dpc material chosen. The more severe the exposure rating, the more important it is to choose the most suitable material for each location.

It has also been the practice to specify 'quality' dpcs only for 'quality' buildings and again it must be emphasised that *it is unwise to cut corners on the weatherproofing of any project.* Even 'temporary' buildings have a habit of becoming permanent. Firms specialising in dpc installations to existing structures are now being asked to replace some of the early bitumen-based dpcs installed prior to the second World War.

4 Detailing of dpcs and related work

4.1 General

As stated briefly in the introduction, some detailing of dpcs is adequately covered in official sources by well thought out details and consequently has been reasonably well understood by most members of the building team (eg horizontal dpcs at the base of walls, **13**). Construction textbooks supplement the 'official' sources, but unfortunately they are usually content to include the same details rather than widen their coverage. Other areas of dpc detailing (eg cavity dp trays) are, however, in most cases ignored or covered by ill-considered illustrations. In CP 121[4], for example, six pages are devoted to chimney flashings compared to two on cavity dp trays for concrete frame and loadbearing brickwork construction. The only publication that begins to approach the subject properly is *Brickwork resistance to rain* by D. Foster RIBA, published by Structural Clay Products Ltd.[16] The AJ *Everyday details*[19] have been very useful in discussing dpc details but without an organised approach to detailing them. (Indeed, an indication of the importance of dpc detailing is given by the fact that the majority of details published in *Everyday details** have a dpc at some point in the construction.) Much 'detailing' of dpcs, eg corners and junctions, has traditionally been left to site agents and bricklayers on site to resolve. In a survey of architects for the original paper on which this book is based, most did not concern themselves with much elaboration beyond the most basic sectional details. The skilled craftsman of the past and the knowledgable architect are not always available today, and both must be more effectively educated in dpc detailing and other constructional matters.

The following sections, therefore, will concentrate on those aspects of dpc detailing which are not well covered elsewhere. In some cases, dpc details cannot be understood unless they are looked at 'in the round', (for too long, working drawings have been exlusively considered in plan and section). While it may sometimes be possible to 'cut corners' in some of these details, eg sealing of laps, this must only be done if considered in relation to the exposure ratings outlined previously.

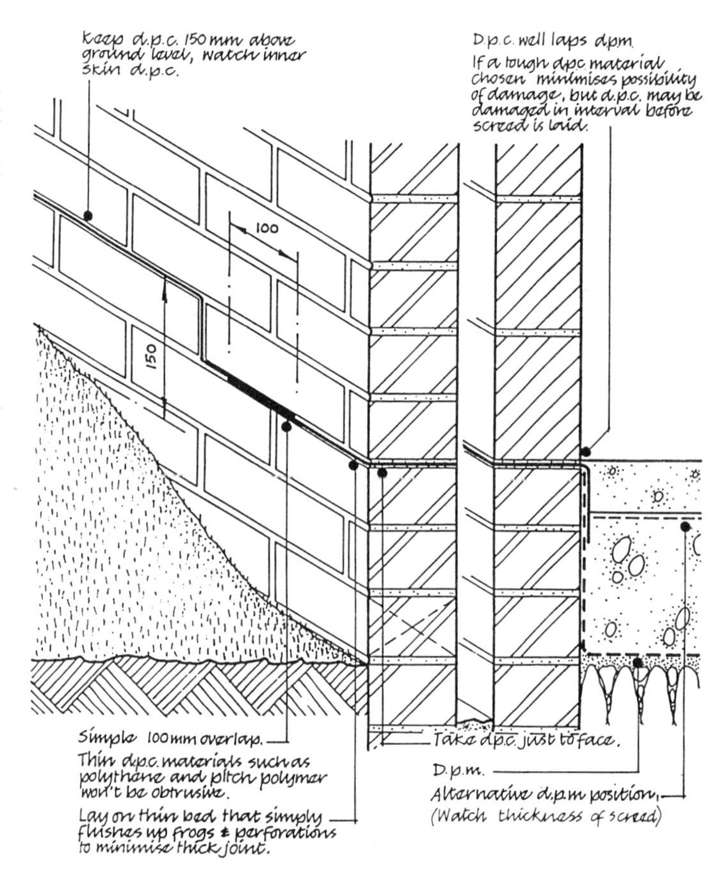

13 *Dpc at base of cavity wall—a comparatively well understood detail.*

4.2 Detailing of dpcs at ground level

At base of external wall

13 shows a typical horizontal dpc at the base of a masonry cavity wall. This detail is basically well understood and therefore not elaborated here; but a number of points are emphasised on the diagram.

This detail usually only fails at some date long after construction when the dpc is bridged by additional soil, new footpath, etc. (For further discussion and althernatives see *Everyday details* 2, 3, 5; for wall dpcs for timber framed buildings see *Everyday details* 6.)[19]

At base of freestanding wall

Free-standing brick walls present difficult dpc problems particularly with regard to lateral stability (because of lack of tensile bond across dpc) and due to the fact that they are exposed from all directions, **14**.

At base of retaining wall

There are even greater problems when the free-standing brick wall becomes a retaining wall, often resulting in long-term efflorescence which is unsightly. Sulphate attack is even more serious and can cause disintegration of the bricks, **15.** Bricks low in soluble salts, and/or sulphate-resisting cement should be used.

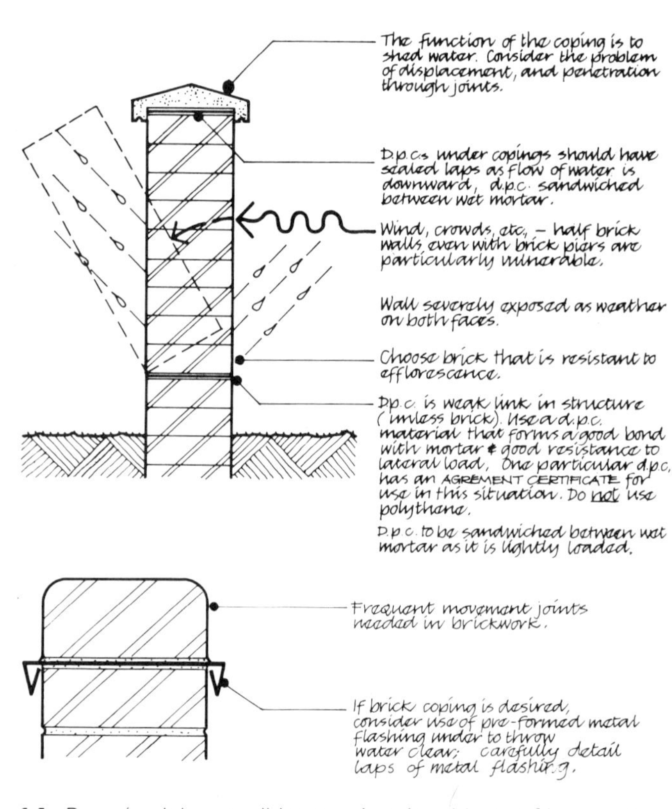

14 *Dpcs (and the possible associated problems of interruption of bond) in free-standing wall.*

* *Everyday details*, edited by Cecil Handisyde and mentioned above is available as a book (The Architectural Press).

General

The current architectural preference for a 'brick aesthetic' in all elements of a building including staircases, planting boxes, etc, **16,** makes for very difficult dpc detailing problems. Many of the most highly-regarded buildings of this kind suffer from serious long-term efflorescence. The efflorescence is caused by the bricks being exposed to very severe wetting, eg a typical staircase detail shown in **17a, b;** other problems are illustrated in **17c, d.** The only real recommendation that can be made with regard to this problem is not to try such a detail except in landscaping situations where water penetration to the underside and efflorescence are less critical. In any case, bricks low in soluble salt content (test to BS 3921) and therefore resistant to efflorescence should be used in these situations; and, because the problem goes further than mere efflorescence in these very exposed situations, even that may not be enough—bricks should always be of 'special' quality.

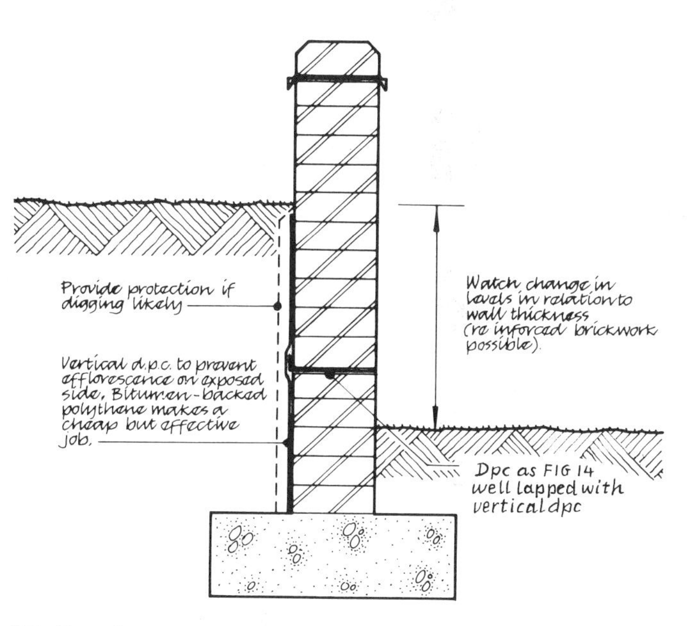

15 *Dpcs in retaining wall.*

4.3 Detailing of dpcs at window openings

General

Dpc detailing around windows, **18,** is reasonably well covered in other sources. It is a very wide subject as the details will vary greatly depending upon the type of window and surrounding construction. It is important that the jamb, head and sill are considered in relation to each other, for it is often the junction of the various dpcs that causes problems. Manufacturers' recommended details should be carefully checked as they often are very careless with regard to dpcs, particularly for steel and aluminium windows fixed directly into brickwork, where tolerances are small.

At window jamb

The detail of the vertical dpc at window jambs will differ considerably depending upon whether the jambs are built flush or rebated and upon the relationship of the window to the face of the wall. (See 'Everyday details' 7 and 8[19] and 'Brickwork: resistance to rain'.[16]) The details shown here **19a, b,** give simple but effective details for timber and steel windows where the windows are built in during the construction of the wall. Where the windows are installed after the walls are completed, much greater effort is required to ensure effective dpcs. (See site installation, section **6.3**.)

16 *The currently popular 'brick aesthetic' creates very difficult problems of dpc detailing (eg at planting boxes on suspended deck); and close attention is required if anticipated appearance is not to be ruined by efflorescence and other failures.*

17a

17a *Another example of the 'brick aesthetic', which can be architecturally very effective, but exposes many elements (such as staircases and parapets) to wind and rain. For typical problem occurring on exposed brick stair, see below.*

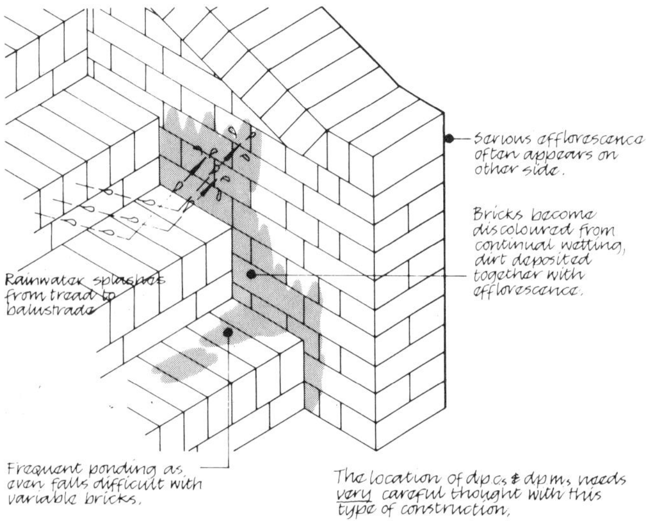

17b

17c

17d

17b *Problems which may occur at exposed staircases. Special quality bricks should always be used in such situations (though not even this will guarantee low efflorescence).*

17c *Though the visual effects of efflorescence above this shop front may be reduced in time, permanent staining and earlier deterioration of the brickwork will almost certainly result.*

17d *Brick on edge copings even with an effective dpc often lead to serious efflorescence. If used at all, bricks should be very carefully selected and only those used that have been shown to be efflorescence free.*

18 *Window surrounds formed of exposed brickwork with recessed joints, and brickwork sills, requiring very careful dpc detailing at top, sides and bottom of opening if water penetration is to be avoided.*

18

Dpc. stapled/nailed to frame before window built in. Important that d.p.c. extends into rebate.

Vertical d.p.c. in one piece, fixed tightly between blockwork and cavity closure, penetrating 25 mm into cavity.

If cavity insulation batts are built-in, it is very important that they are behind the V.D.P.C. with a good lap to the joint (some p.v.c. cavity closer/d.p.c.s work well with cavity insulation.)

If cavity insulation installed, the external skin cannot be returned to form the cavity closer as insulation on wrong side of V.D.P.C.

6 mm joint pointed with mastic (quality of mastic depending upon exposure, height of building etc.)

19a

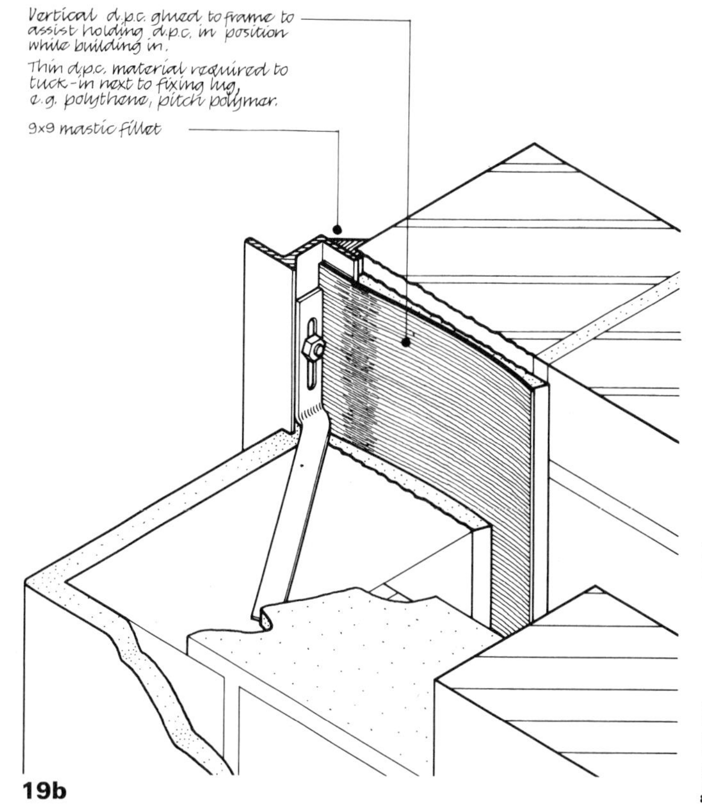

19b

19a, b Vertical dpcs to timber and steel window jambs respectively. Both assume windows to be built in during wall construction (problem of effective dpcs becomes more difficult if windows are installed after completion of walls). In a later section ('Installation') the full assembly of lintel, jamb and sill dpcs will be shown fitted together to form a continuous water barrier.

At window heads

Again, the detail of the cavity dp tray at window heads varies greatly and only two alternatives are give here, **19c, d.** See also 'Everyday details' 8 and 10.[19] Pressed metal lintels in most cases provide an integral cavity dp tray, although some types are easily bridged by mortar. For further details, see AJ 'Products in Practice'; sills and lintels 10 Oct. 1979.

At window sills

The function of a dpc at the window sill, **20,** is two-fold; first, it must (in the case of timber sills particularly) stop damp rising from rain-saturated wall below; and second, it must prevent water penetrating any joints in the sill getting to the inner skin—particularly with brick sills. Although many sills are fairly impervious themselves (eg pre-cast concrete) a sill dpc should still be provided as a precautionary measure as sills are subjected to very heavy run-off of water. The ends of sills are important risks, and particular thought must be given to what happens at these points, **20.**

4.4 Detailing of dpcs in parapets

General

The detailing of dpcs in parapets is critical because parapets are severely exposed to the elements in all but the most sheltered locations. A good number of reported dpc failures in buildings recently have been concerned with parapet dpcs. Indeed, a famous American architect is reported to have said to an assistant, 'you may detail a parapet wall as long as it is preceded by your letter of resignation'. Even traditional parapet details suffer water and efflorescence staining, **20a.**

Low and high parapets

Suggested details for low and high masonry parapets in cavity brickwork are shown in **21a** and **b.** Solid parapets will almost certainly lead to problems of water penetration and are not recommended unless very great care is taken to get the detailing right: because of the riskiness of solid parapets they are not detailed here; but see 'Everyday details' 21.[19]

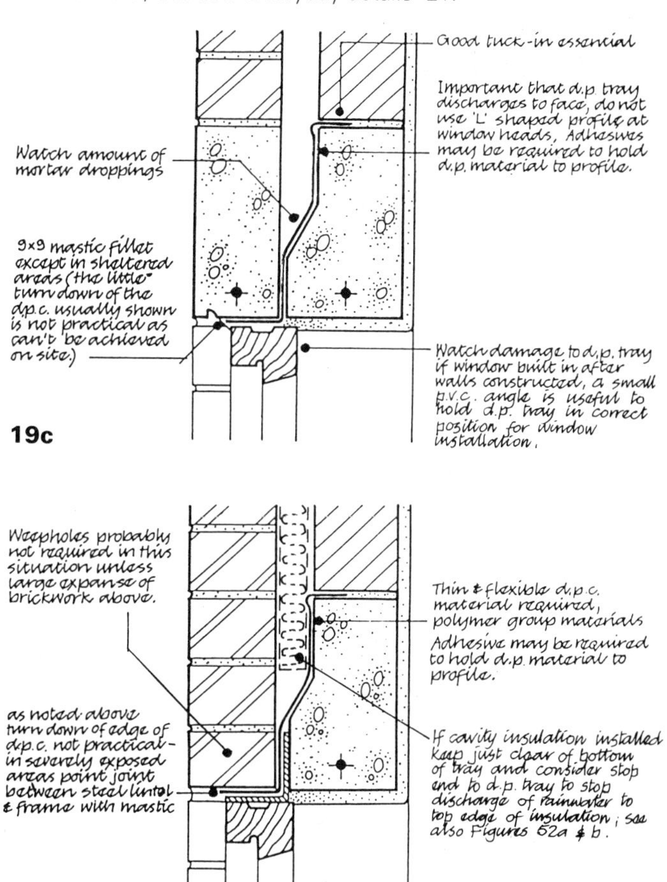

19c, d *Two of many alternative cavity dp tray details at window heads.* **c** *shows reinforced concrete or brickwork lintel* **d** *shows metal angle.*

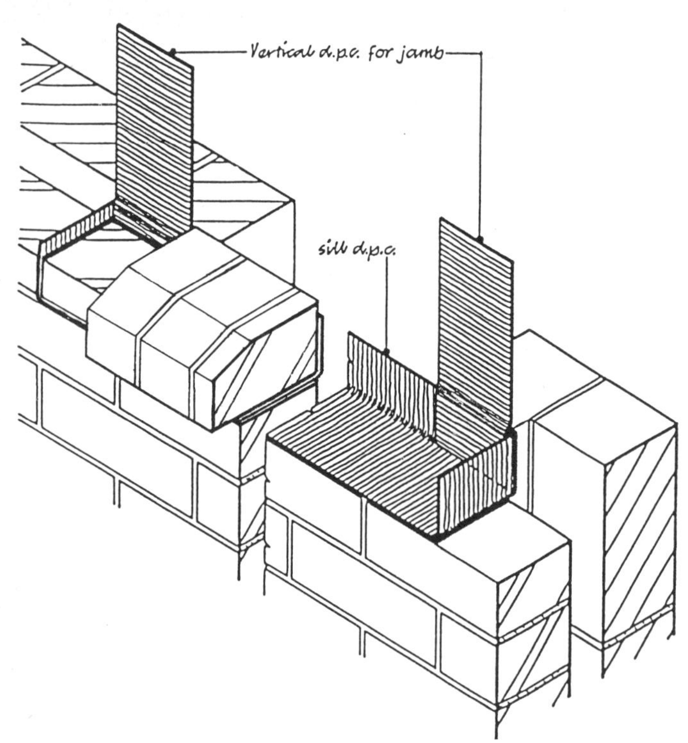

20 *Cutaway section showing how vertical dpcs at window jambs overlap, and discharge onto, sill dpc to form watertight assembly. Sill dpc is extended one half-brick beyond each end of sill to link up with vertical dpc down cavity closer face; and is turned up at each end to height of one brick course, to form tray. Entire assembly of lintel, jamb and sill dpcs is shown in later section ('installation'). See also AJ 10 Oct. 1979 'Products in Practice'.*

20a *Traditional parapets suffer water staining and subsequent efflorescence staining and in the long term, loss of mortar from the joints of the coping stones leads to an uneven drip pattern under open joints, as in centre of picture.*

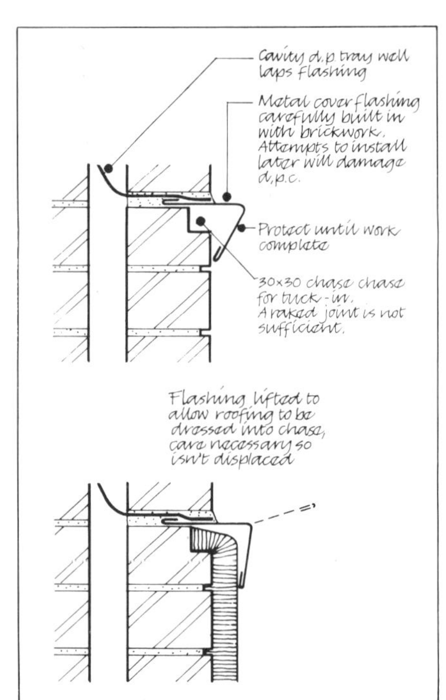

22

21a, b *Dpcs in, respectively, low and high parapet walls, both in cavity construction. Solid parapets are not recommended.*
22 *Detail at cover flashing. Two stages of installation.*
23 *Abutment of pitched roof to external cavity wall.*

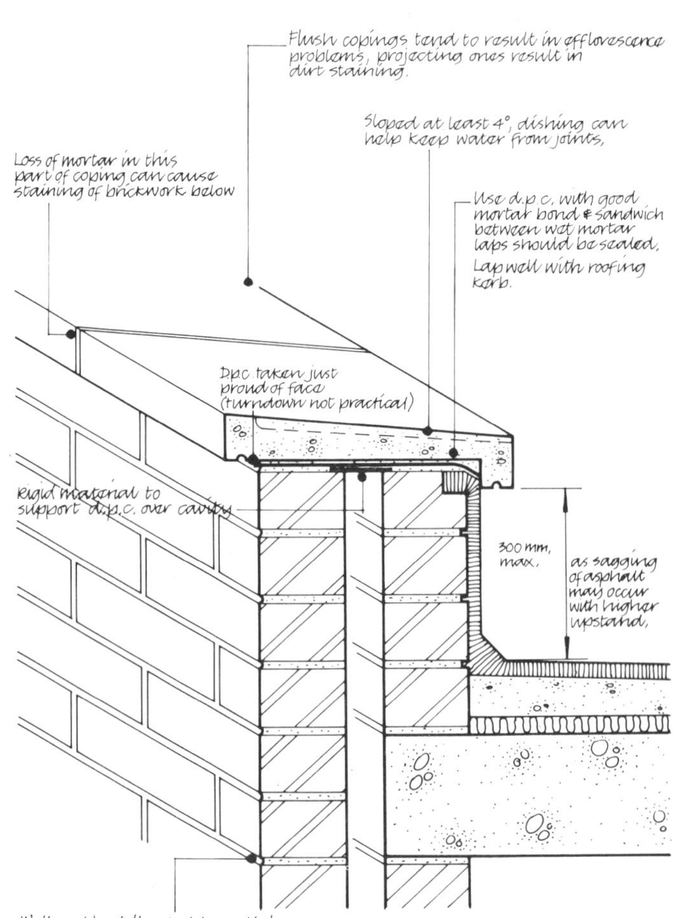

21a

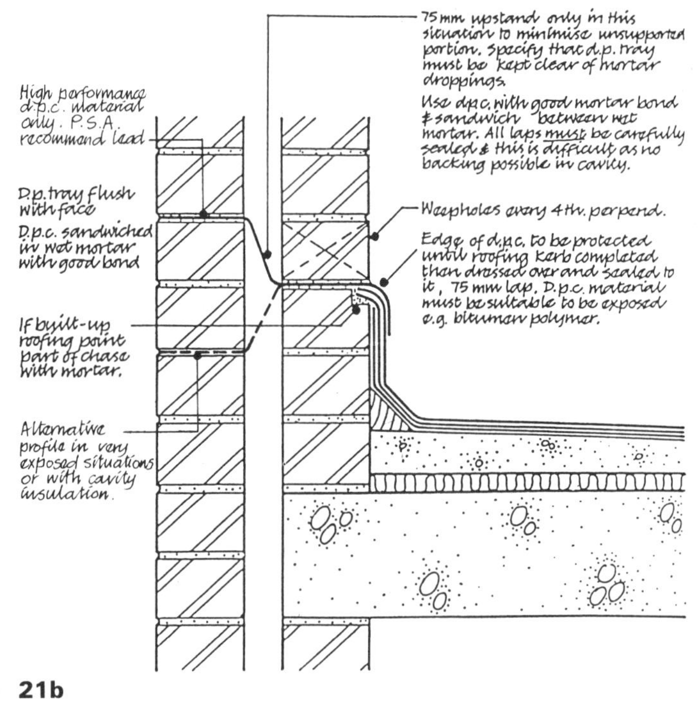

21b

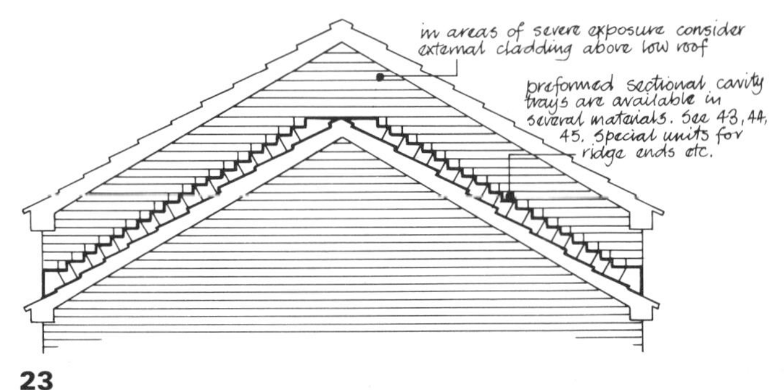

23

The junction between the cavity dp tray and the roofing kerb is a tricky one with many different suggestions seen in the published sources. With some of the tough, new dpc materials available it may not now be necessary in some cases to insert a separate cover flashing with all its attendant problems, **22.** Temporary protection of material is essential in both cases.

Roof/wall abutments

The abutment of a pitched roof to an external cavity wall extending to a higher level makes great demands upon the designer and the site operative. It is usual to detail a cavity dp tray in short legths dressed over a lead soaker (see 'Everyday details' 16[19]). Forming short lengths of cavity dp tray and lead flashings is difficult and time-consuming; several manufacturers now make preformed units which can be installed much more quickly and with less reliance on extra-careful workmanship on site, **23.** Alternatively, the portion of wall above the roof can be made waterproof by external cladding. The detailing of a flat roof abutting an external wall is much simpler and similar to the parapet detail shown previously, **21b,** see also **31.**

4.5 Detailing of cavity dp trays in framed and loadbearing buildings

General

The detailing of cavity dp trays has been largely ignored in official sources and textbooks although they are often necessary in modern forms of construction. The author has made a particularly thorough study of cavity trays based upon both site experience and discussion with others in the building team. Some of the comments on mortar bedding, weepholes, etc that follow also relate back to other types of dpc detailing.

Correct profile

The traditional cavity dp tray was usually shown as a 'Z' shape with a 75 mm upstand as included in CP 121:101:1951, **24.** This detail is now superseded but still often seen on working drawings. The detail was worked out at a time when malleable metals were used more often than now; when mortar haunching behind the tray could be assumed to be included; and when lifting battens were commonly used to keep the cavity clean. It is an inadequate detail today because flexible dpc materials are most often used, mortar haunching is seldom employed due to costs, and it is virtually impossible to get bricklayers to keep the cavities clean. Consequently, there is the risk of damage to the unsupported section of the dpc when cleaning out the cavity after completion of the wall, or if the droppings are left the 75 mm upstand may be bridged. The detail in the 1973 Code of Practice, **25,** goes some way towards overcoming these criticisms in that it increases the upstand to 150 mm, but it is virtually impossible to achieve this profile in practice and much of the tray is still unsupported.

The present author would suggest that the best profile is an 'L' shape as shown in **26.** This allows for a maximum amount of droppings without bridging; easier removal of those droppings without damage; and maximum support for the dpc without haunching. Most materials can be formed into an 'L' shape more permanently than the 'Z' shape commonly shown on drawings but rarely achieved on site. As will be explained later, the 'L' shape also makes the formation of corners and junctions much easier. On the other hand, it should be borne in mind that a slightly wider strip of material will be required for the 'L' profile detail. The author would suggest (contrary to some current Code of Practice details) that a minimum 150 mm upstand is provided at all locations (except parapets) using a cavity dp tray—eg floor levesl, window heads, etc. In some cases a 225 mm upstand will be required to suit blockwork coursing, but the upstand should not be increased above 225 mm as it may easily become displaced, though it may sometimes be possible to fix large dp trays to the substrate with mastic or adhesive

Termination of cavity dp tray

There is considerable controversy within the architectural profession and elsewhere on whether the termination of the dp tray (indeed any dpc that is meant to discharge rainwater) should project, be flush or slightly recessed, **27.** The slightly

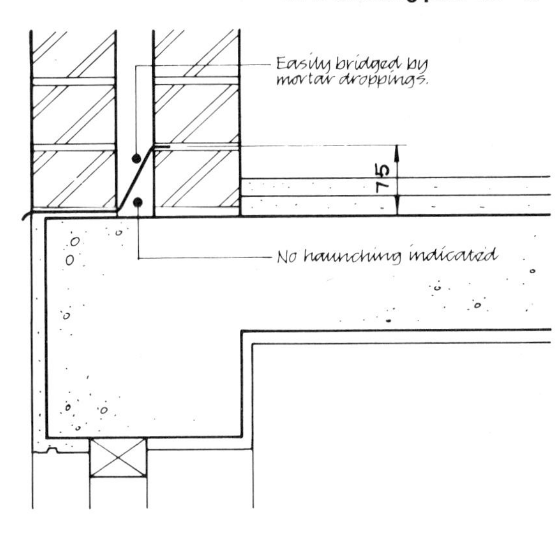

24

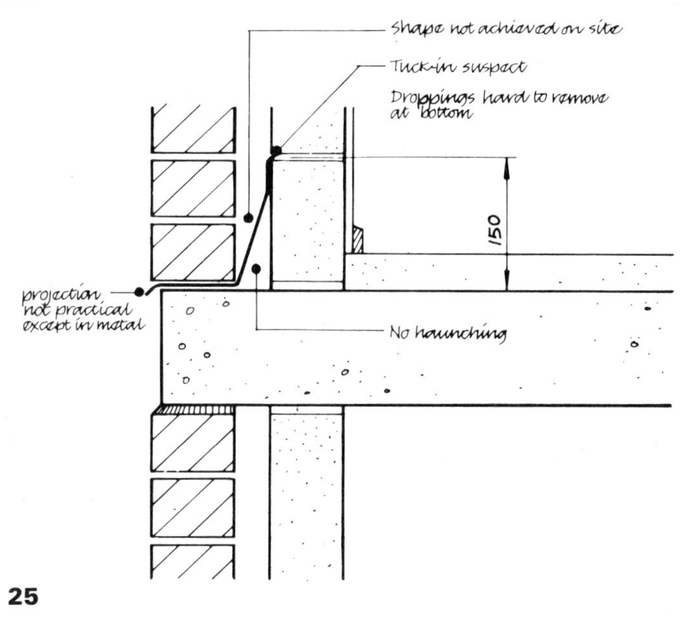

25

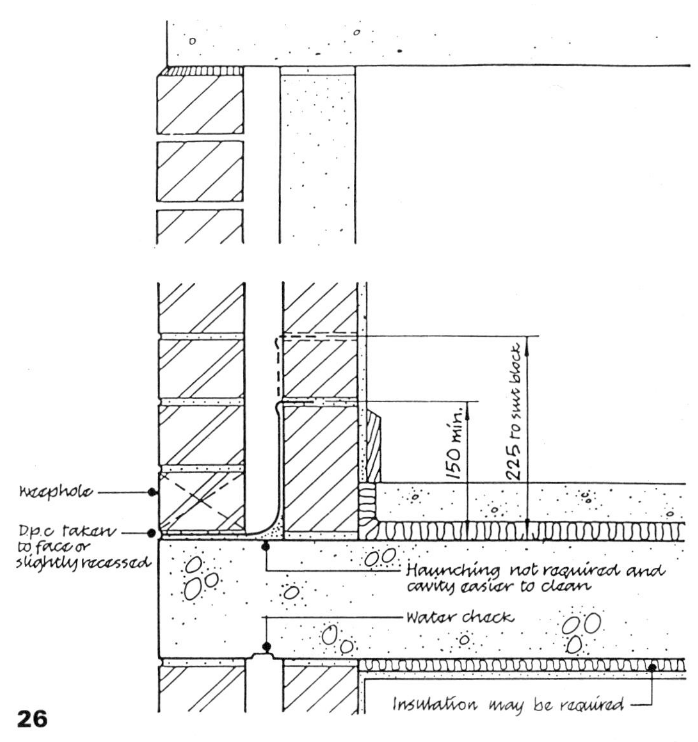

26

24 to **26** *Three alternative basic profiles of stepped dpcs.* **24** *shows Z profile from CP 121:101:1951 (page 46, fig 6), now superseded but still appearing on architects' drawings.* **25** *shows the updated version from CP 121:1973 (page 61, fig 4); and* **26** *shows the author's suggested detail, now seen in many other publications.*

a PROJECTING **b** FLUSH **c** RECESS

27

turned down projection usually shown theoretically acts as a drip for water discharged by the tray and masks the joint under the dp tray. The projection undoubtedly derived from earlier flashing details using malleable metals. The unsightliness of bitumen oozing from this projection when bitumen-based dpcs came into use, however, led architects to specify that the dpc be slightly recessed for facing work. (There is a standard clause in the National Building Specification[23] to this effect). It is virtually impossible to achieve the neat little projection with the flexible dpcs most commonly used today. The material will project out flat and easily become damaged, **28.** The only way to achieve a real drip to the dp tray is to make a separate piece in metal tucked under the edge of the dpc, **29,** but this is costly to install and easily becomes displaced.

In the author's survey of architects regarding dpc work only 10 per cent detailed a projecting dp tray. From the survey there appeared to be no connection between flush or slightly recessed dp trays and water penetration. I favour bringing the edge of the dpc flush with the face in most cases. Bitumen-based dpcs are fairly compressible and any significant set-back may in certain instances lead to spalling bricks, **30** (one valid factor in favour of flush or projecting dpcs).

Haunching behind cavity dp tray

As stated earlier, haunching is seldom specified today although it may be useful in certain locations where the sole of a dp tray would otherwise be unsupported, **31**. In any case the haunching will not ensure a completely reliable backing unless the dpc material is fixed to it by adhesive or other means. The haunching recommended in 'Everyday details' 15[19] cannot be achieved consistently on site.

Mortar bed to cavity dp trays

The use of a mortar bed is mored common than haunching, but it is by no means universal and should be examined in the light of current practice. The original function of a mortar bed was to flush up brickwork underneath the dpc to provide an even surface free of projections, that might damage the dpc material. In official sources there is confusion as to whether the dpc is to be bedded onto the mortar while still wet or be allowed to dry. It is certainly difficult to neatly lay many of the dpc materials onto a wet mortar bed (eg bitumen/lead in cold weather) but many current Agrément certificates on polythene and pitch polymer state that these materials should be sandwiched between wet mortar.

This requirement is undoubtedly the result of problems arising on a few sites where it was alleged that rainwater tracked underneath dp trays that were bedded directly onto concrete floors, etc, **32.** Laboratory tests have shown that water can be forced under some dp trays with constantly high pressure differentials, but site evidence indicates that with the naturally fluctuating air pressure, this does not occur to a significant degree. However, if the surface onto which the dp tray is laid is very rough or grooved, these defects could allow water to be forced underneath it in significant quantities.

The designer must decide in each situation if he feels a wet or dry mortar bed is necessary but in some cases it will not be. Some form of support is needed especially under sealed laps, which are difficult to seal otherwise. A possible alternative to a mortar bed is to gun a mastic bead onto the base on which the dpc is then bedded. *Brickwork: resistance to rain*[16] suggests incorporating a water bar behind the sole of the cavity tray but this would be very difficult to incorporate on site and it seems doubtful whether it could be made effective. Locations where there is a need for resistance to lateral movement (eg garden

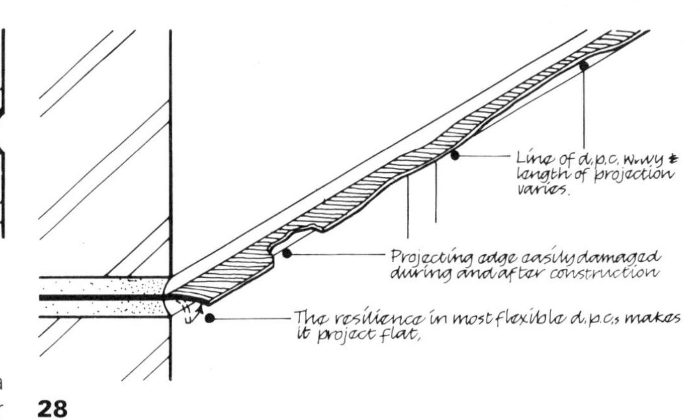

28

27 *Shows three alternative edge terminations of dp trays as designed; and* **28** *the termination of a 'projecting' dpc as usually found on site: wavy, irregular, and damaged.*

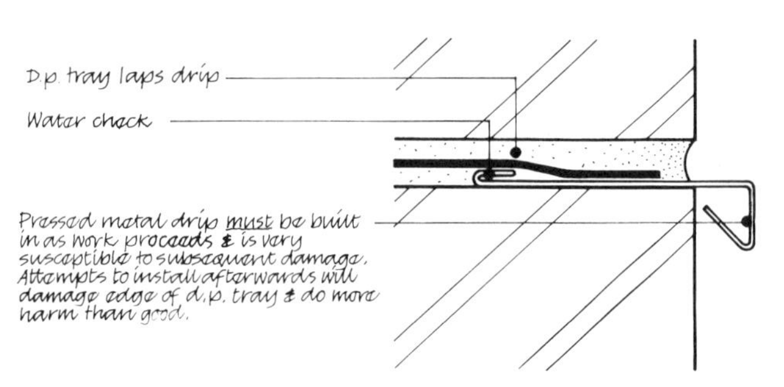

29 *The only way of providing a real drip to a dp tray: instead of* **27a** *(virtually impossible with flexible dpcs), a separate metal drip is tucked in under the dpc edge.*

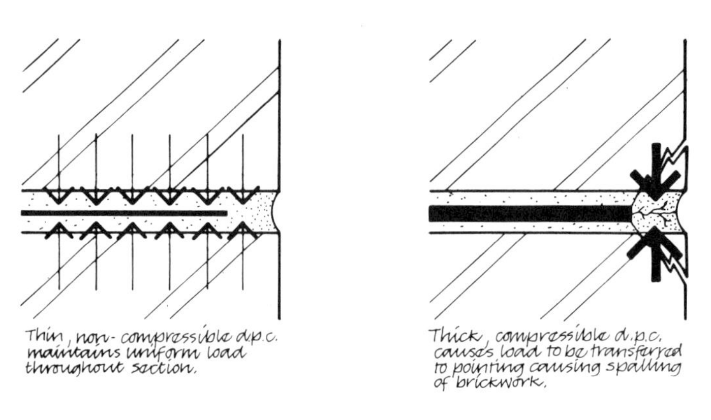

30 *Problems in brickwork with compressible dpc and recessed termination.*

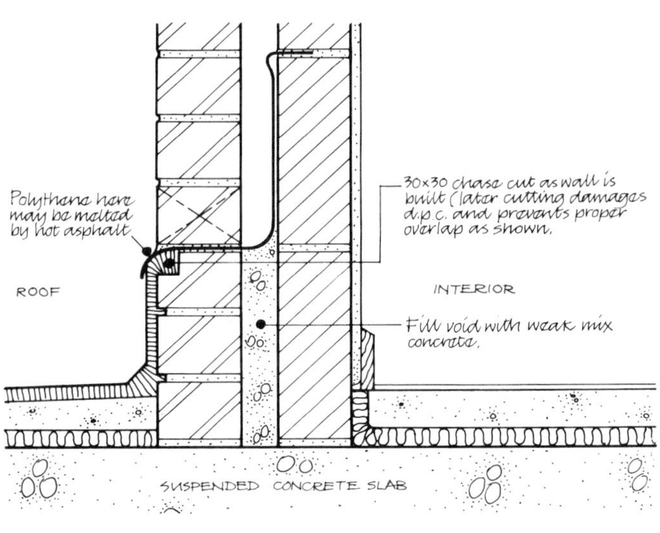

31 *Haunching to small void under cavity dp tray.*

walls and parapet copings) should certainly sandwich a slip-resistant dpc between wet mortar (eg brick or slate dpc, or 'Permagrip' or similar product), in addition to the wall being designed to be stable from the dpc upward in case of failure.

The provision of weepholes

Weepholes are normally provided at the base of cavity walls and directly above any cavity dp tray or flashing that bridges the cavity to allow water penetrating into the cavity to drain away. The weepholes are usually the full height of one brick course (brick and joint) spaced in every fourth perpend joint. For visual and other reasons they are sometimes reduced in size down to a 9 mm diameter hole, **33.** Site experience indicates that the size of weepholes is not critical as long as they are not blocked in the cavity with mortar droppings, etc. The spacing of weepholes within panels of brickwork can be important and they should be set out starting two bricks from columns and other vertical obstructions, **34.**

There is a considerable body of experienced opinion, however, which considers (though there is little formal evidence to refer to) that weepholes create more problems than they resolve. On very exposed sites, wind pressure has been noted forcing rain into the cavity through the weepholes.

Some designers have, therefore, omitted them, and there are a number of high-rise load-bearing brickwork developments in the UK where weepholes have not been specified or built into the work. In most cases, it has been claimed, no problems with water penetration have been experienced; in these cases, the water that gets into the cavity must drain out through the bottom mortar bed joint. On the other hand, it is argued by an opposed body of opinion that the head of water that may be formed in a cavity by lack of weepholes, could be dangerous; and that even very minor damage to cavity trays would become critical in these circumstances.

To deal with the problem of blowback, some expert opinion has for a number of years recommended the use of drainage tubes placed where the weepholes are normally located, for severely exposed sites. It is thought that these tubes prevent blowback and encourage draining of the cavity. They can be 'L'-shaped 'weeptubes' or 'T'-shaped 'venturi tubes', **35.** The tubes are placed into preformed holes after ensuring that the holes are not blocked with mortar droppings.

Site monitoring on exposed buildings could establish how weepholes and weep tubes really function, and indeed, whether weepholes or tubes are necessary at all. Meanwhile, it is safest to provide some form of drainage to cavities.

4.6 Lapping and sealing of dpcs

It is necessary at the detailing stage to consider the lapping (and sealing) of dpcs carefully. The lapping of dpcs is very badly covered in Codes of Practice and British Standards, although more detail is given in BRS Digest 77[9] and a publication by the Brick Development Association, Practical Note 6, *Damp-proofing courses and flashings with brickwork and blockwork*[3]. However, some of the recommendations regarding lapping and the sealing of laps in these latter publications are not easily achieved on site.

There are two basic types of laps required:

- For rising damp, **1a,** most dpc materials should simply be *overlapped* about 100 mm to 150 mm.
- For downward, **1b,** and horizontally penetrating damp, **1c,** (usually rainwater) all dpc materials should be *lapped* and (if dpc is horizontal) also *sealed.*

The sealing of laps in dpc materials has been omitted in many (if not most) cases in the past, but for buildings of moderate to severe exposure this practice will lead to trouble. **Table VI** gives recommendations for sealing laps in the various dpc materials, which are at variance with the 'official' publications but which are practicable on site and effective in producing a good watertight seal. Particular mention should be made of the difficulty in sealing laps in polythene dpc material, and serious consideration should be given to specifying an alternative dpc material in cases where there are many laps that require effective sealing. Manufacturer's knowledge regarding the sealing of laps and its attendant problems is in many cases sadly lacking.

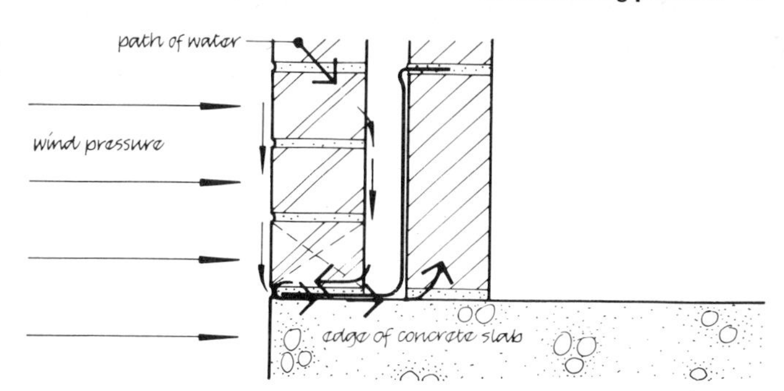

32 *Possible path of water tracking under unbedded dpc (particularly if upper surface of concrete bed is very uneven).*

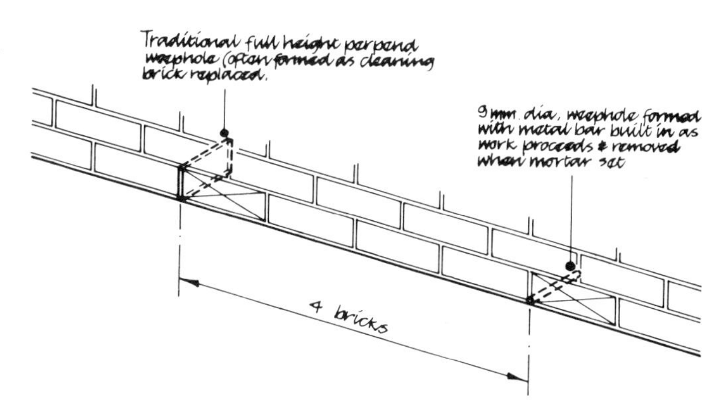

33 *Formation of weepholes.*

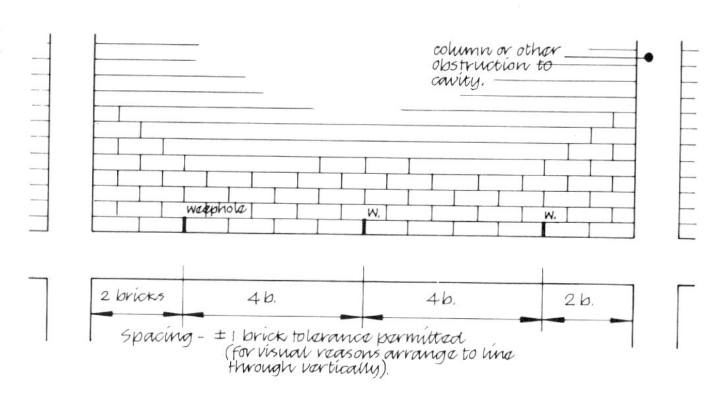

34 *Setting-out of weepholes in panels of brickwork. A contractor may try to get away with only two weepholes in a situation such as this. See specification clauses p25.*

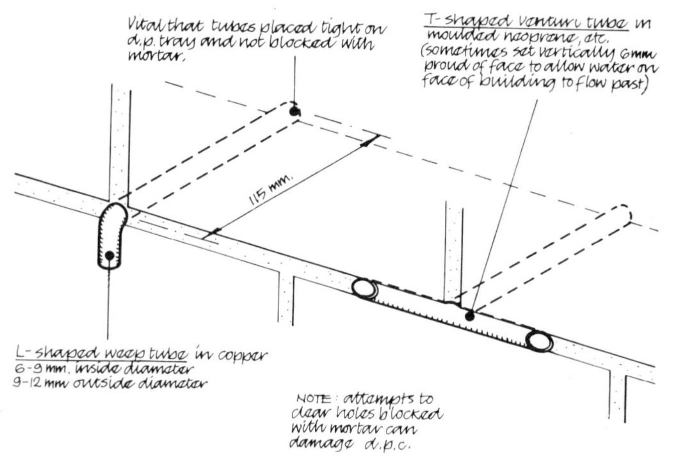

35 *Typical types of tube for draining cavities in severely exposed areas (preventing blowback of rainwater into cavity).*

a

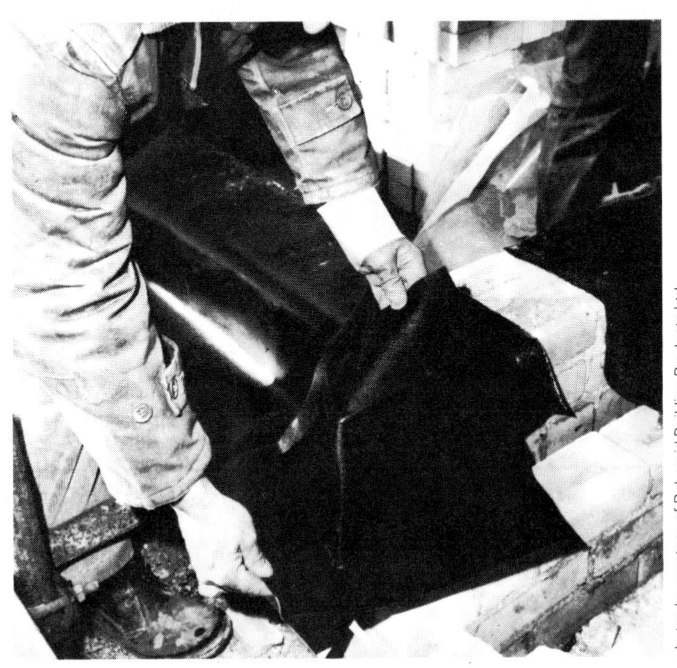

b

photos by courtesy of Rubcroid Building Products Ltd

36 *Architects tend to leave awkward corners and junctions such as these to be solved on site, instead of solving them on the drawing board.* **a** *Preformed pitch polymer cloaks used to cover dpc joints at changes of direction.* **b** *Fit of cloak being checked before installation. Actual installation will be done after bed of mortar has been laid on brickwork surfaces. Note that the natural shape of the dpc away from the preformed corners is an L profile. Sequence of installation would be similar with preformed polypropylene.*

Table VI The sealing of laps to prevent downward flow of water

Material	Method
Malleable metals (except copper)	100 mm lap well bedded in bitumen-based mastic after priming metal surfaces[1].
Bitumen based	Soften 'free' bitumen in two pieces to be joined with blow lamp, and bed together. A few types do not have sufficient bitumen to do this, therefore bed these in bitumen-based mastic.
Polythene	Use double-sided adhesive tape to manufacturer's recommendation. Priming may be necessary. Ensure seal is permanently held in place by construction, otherwise lap may 'spring' apart in time[2].
Polypropylene	Sealed with mastic sealing tape, either factory-fitted or placed in situ.
Pitch/bitumen Polymer	Use adhesive or mastic to manufacturer's recommendations. (In some cases, double-sided tape can also be used as recommended by manufacturer).

Notes:
1 Welted joints usually recommended but difficult to achieve on site except in copper, and are very bulky.
2 Welted joints usually recommended but impossible to achieve on site.
In many cases manufacturer's instructions are inadequate and in these cases a method should be selected based upon the above general instructions for the type and tests should be carried out on the effectiveness of the seal, compatibility of materials, etc.
See section 6.3 for diagrams.

4.7 Corners and junctions in cavity trays

Left to site expertise

Corners, junctions, etc, **36,** have traditionally been left to the bricklayer on site to resolve, sometimes in consultation with the Clerk of Works for the more complicated situations. Very few architects have given any consideration to the matter. It may not any longer be sufficient to rely on site expertise, as the standard of knowledge and workmanship may not be of sufficiently high quality. In addition, labour-only sub-contracting and bonus schemes tend to ignore the 'incidentals' of bricklaying, and there is little incentive, therefore, for the bricklayer to do a careful job in forming corners and junctions.

Not covered in codes and standards

Although it is hard to believe, no official document or textbook gives any consideration to corners and junctions in cavity dp trays. *Brickwork: resistance to rain*[16] by D. Foster is the first publication to consider the problem in any detail.

Corners and junctions have been the cause of a number of cases of water penetration, particularly with medium and high-rise load-bearing brickwork in exposed situations. Typical problems have concerned external and internal corners, junctions with doors and windows and brick infill panels to concrete frames. Corners and junctions in dp trays can only be effectively considered in three dimensions, and isometric sketches are very useful in this regard.

Typical details to be considered

Before commencing the detailing of 'difficult' dpc junctions, the designer has to systematically go through the design as a whole and pinpoint the many situations which require careful analysis and explicit detailing. For instance, **37** shows a cavity dp tray and it is obvious that there is more to it than the usual simple sectional details (at points A, B and C) on architects' drawings will indicate. To make a really effective dp tray, the material must be correctly cut, lapped and sealed at the external ('salient') and internal ('re-entrant') corners; at the junction; and at the change in level.

A second instance of a junction often overlooked is an external door opening onto an access balcony, **38.**

A third is the intersection between perimeter column and brick infill panel, in brick infill/concrete frame construction. The relationship between the cavity of the infill panel, and the concrete column, **39,** greatly affects the complexity of the cavity dp tray formation.

A fourth instance is the complex type of junction which often occurs in precast panel construction; a typical example of the dpc flashing which may be required is shown in **40.**

Formation of corners, junctions etc

Having highlighted the various corners, junctions, etc, in the cavity dp trays, the designer should decide on the way they are to be formed. In the past they have almost invariably been formed in situ by cutting, lapping and sealing. If the project is a small one, or one with many different situations, this will undoubtedly be the only feasible choice. It is possible with careful workmanship by the bricklayer and close attention by the contractor's supervisory staff to form them perfectly satisfactory in situ. (See the later section 'Site installation', section 6.3 ('Method of forming corners and junctions on site') for details.

If it is a large repetitive development, however, the designer may decide that *preformed* components are desirable. Corners, junctions, and so on can be prefabricated in most dpc materials but very few manufacturers are really able to produce them at

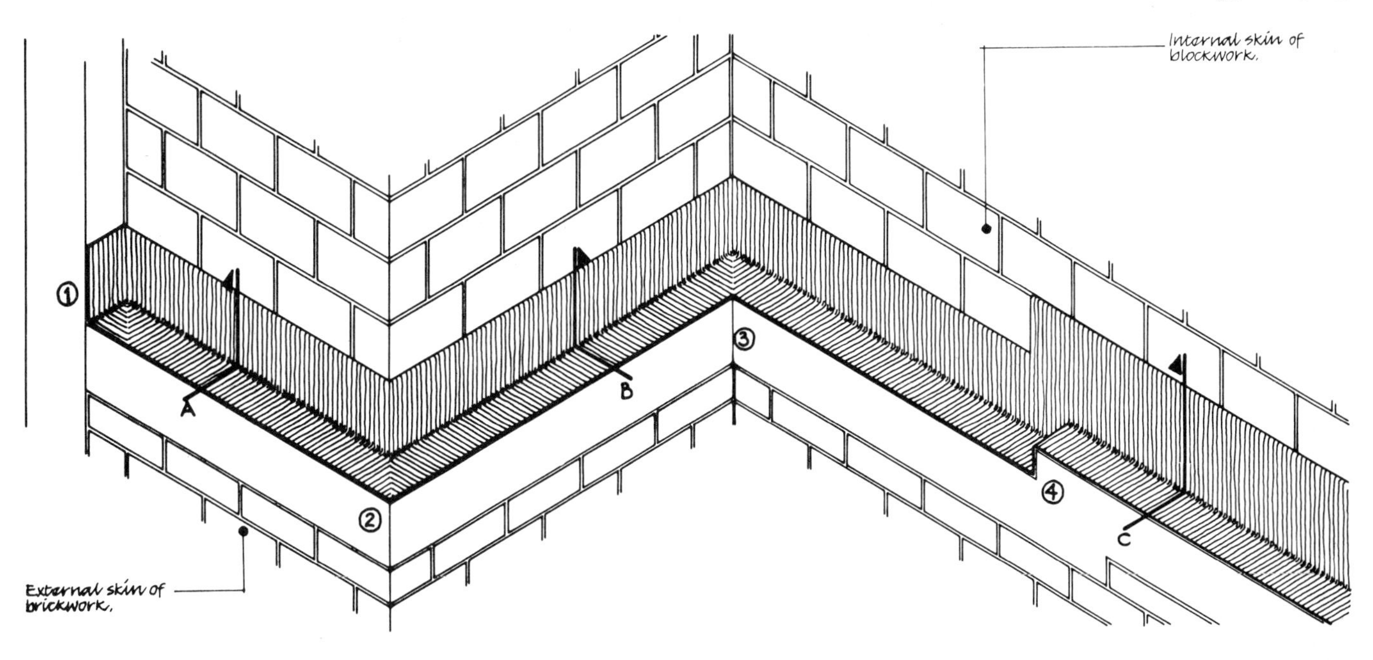

37 *Isometric sketch of typical cavity dp tray (external leaf of brickwork above dpc omitted for clarity). Architects' drawings frequently show simple sections at A, B and C; and ignore the awkward junctions* 1, 2, 3 *and* 4.

present (see **Table III**). The market for preformed dpc components is, in fact, largely dominated by one manufacturer who produces pitch polymer 'cloaks' to standard profiles by a welding process, **42.** The cloaks and the dpc material are covered by an Agrément certificate. These standard cloaks are often not suitable for a particular project and most of his output consists of 'specials' made to order, such as would be required for the complicated junction shown previously in **40.** The standard pitch polymer cloaks are to a 'Z' profile, **41a** and the author would suggest that 'L' profile cloaks as suggested in **41b, 41c,** would maintain the advantages of the 'L' shaped dp tray discussed in section 4.5. Pitch polymer 'cloaks' are sometimes used in combination with another dpc material for the straight runs, eg polythene. If this is done, great care must be exercised at the laps to ensure that the two different materials are joined effectively.

Part of the range of commercially available preformed cloaks in pitch polymer is indicated in **43a–d,** including a 'special' developed by the manufacturer, in collaboration with the architect, for a complex recurring junction in a housing project. Preformed trays are now available in polypropylene and some are shown in **43e, f**: note the L-shaped profile. Using these components, assemblies can be made up for the complete window surround, or for complex ring beams. Typical units in galvanised, pressed or stainless steel or lead are shown in **44.** If such attention to detail (and collaboration between architects and dpc manufacturers) were more common, there would be fewer cases of damp penetration; and the currently favoured 'brick aesthetic' would be less frequently ruined by efflorescence, staining, and other types of damp-associated failure.

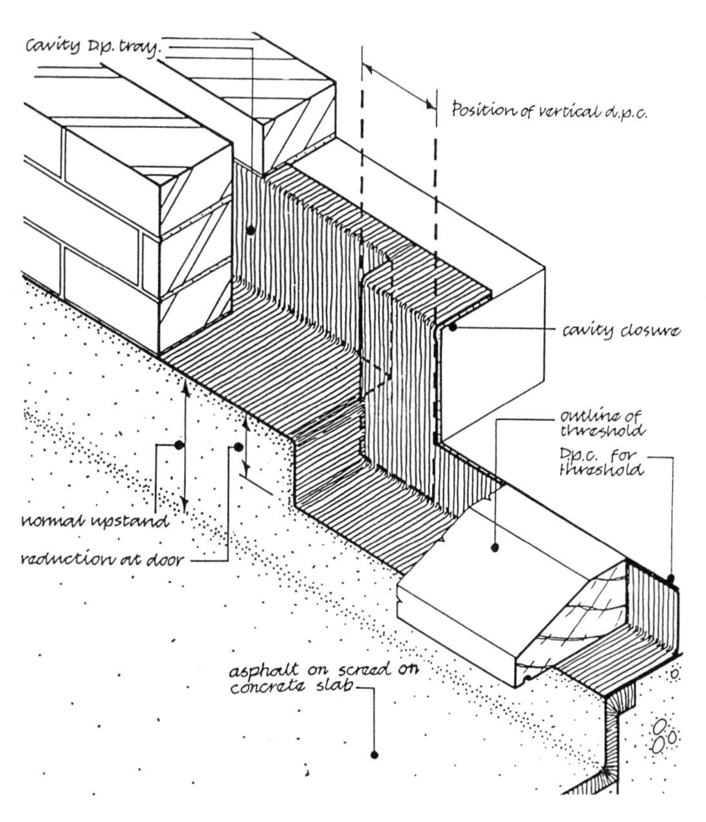

38 *Another frequently overlooked junction: the meeting point between a straight run of cavity dp tray, and the threshold dpc where an external door opens onto a balcony. Complex cutting, lapping and sealing is required to form watertight assembly; and vertical dpc which fits into jamb of door must link up properly with threshold dpc. Special preformed cloak may be designed in consultation with manufacturer for repetitive situations (see* **43d***).*

4.8 Cavity dp trays to brickwork supporting angles

As discussed earlier, there has been concern in the construction industry about the stability of brick cladding to concrete framed buildings. The design of buildings so that they accommodate differential movement (shrinking frame, expanding bricks, deflection, thermal movement) has become more sophisticated. Stainless steel angles are now often required to support sections of brickwork, and as they bridge the cavity, cavity dp trays are required. The detailing and workmanship needed are very exacting for all those involved in the building process. A typical situation is shown in **45** and highlights important matters to consider. Special sliding anchors to stabilise the wall against wind loading often bridge the cavity and may cause problems with water penetration.

The author believes that in many cases the angles and stability ties that are being introduced will cause more problems than they solve: particularly in more cases of water penetration and a greater likelihood of unsightly cracks in blockwork and plaster. The common use of mastic to point the joint will certainly lead

Cavity continuous
Vertical flap must be fixed to column, either by adhesive (as shown) or by tucking into groove.
If floor levels not expressed, cavity continuous horizontally and vertically
vertical d.p.c. necessary.
Partial stop end to cavity must continue around column.
water check grooves.
Full stop end to tray.

39 *Effect of relationship between concrete frame, and position of brick infill panel, upon dp trays. Brickwork omitted above dp tray for clarity. Trays are shown to the L-profile recommended previously (see* **26** *and compare* **41c** *and* **60***).*

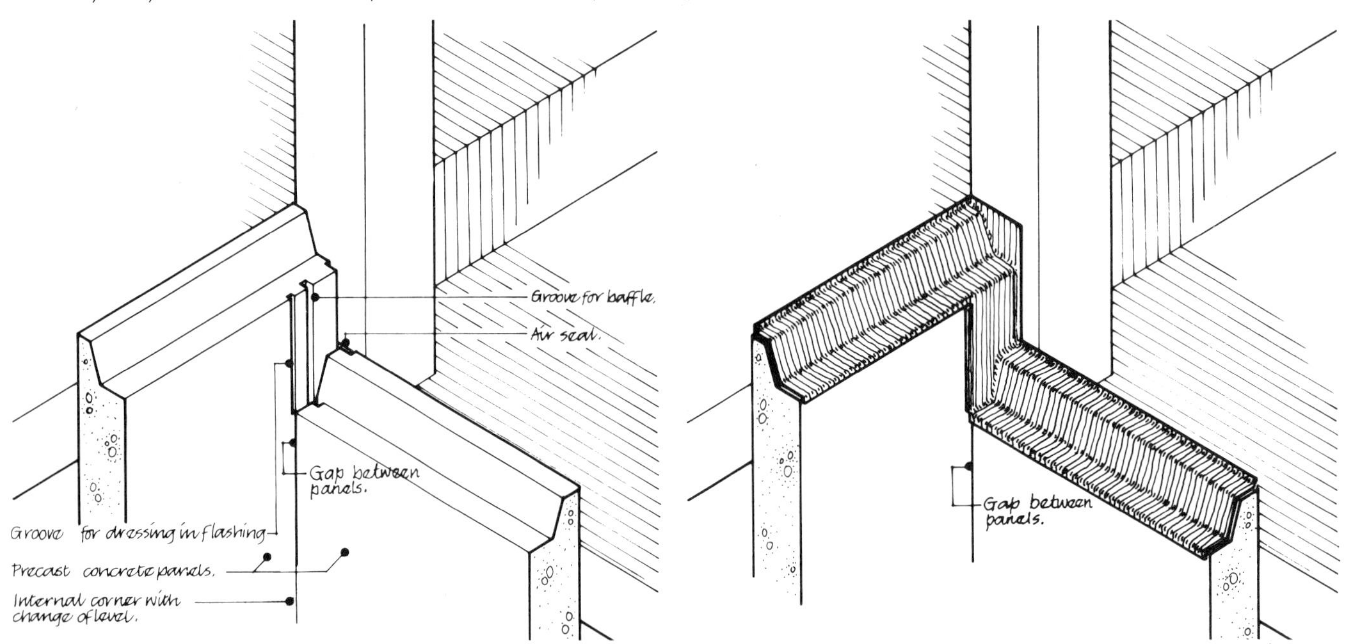

40 *Precast concrete panel construction can require very complex flashings, and preformed components may be necessary to ensure adequate standard of workmanship.*

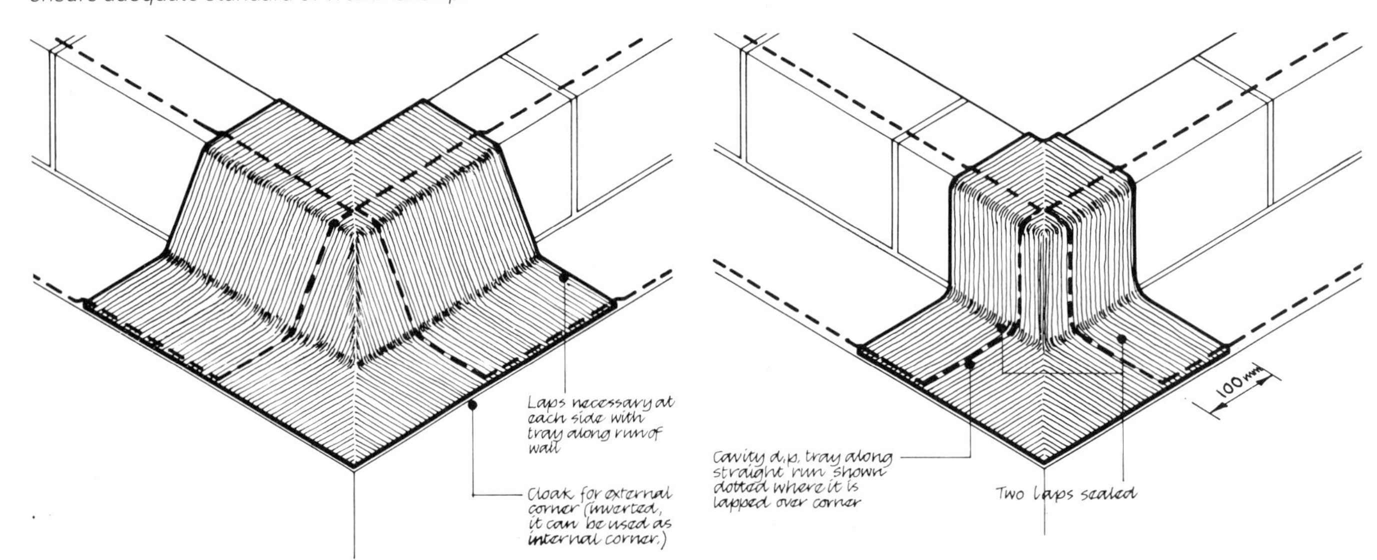

41a *Standard Z-profile cloak for external corner, as currently available. Straight runs shown in broken line.*

41b *Proposed standard L-profile cloak for external corner; this would allow precise matching up with L-profile straight runs, as recommended by author. Inverted for internal corner.*

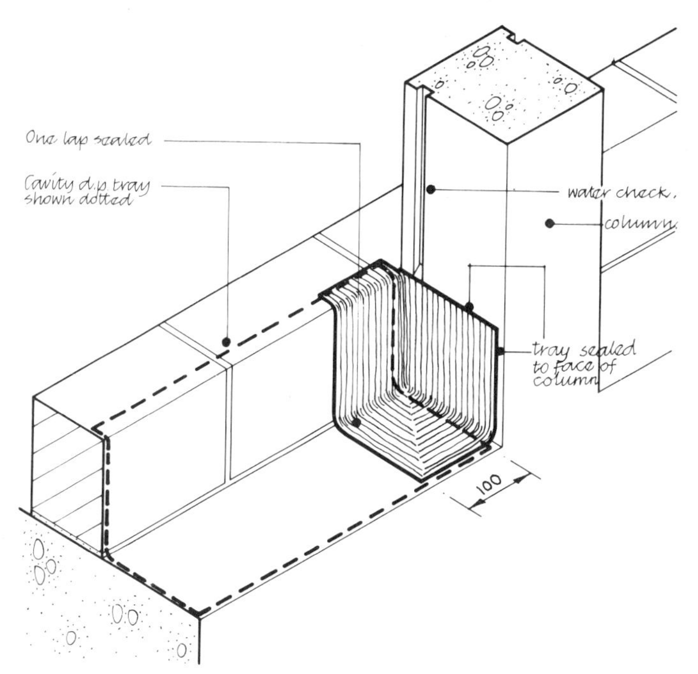

41c *Proposed standard L-profile cloak for dp tray stop-end.*

42 *Standard Z-profile cloak for change of level, as currently available (see* **43b***).*

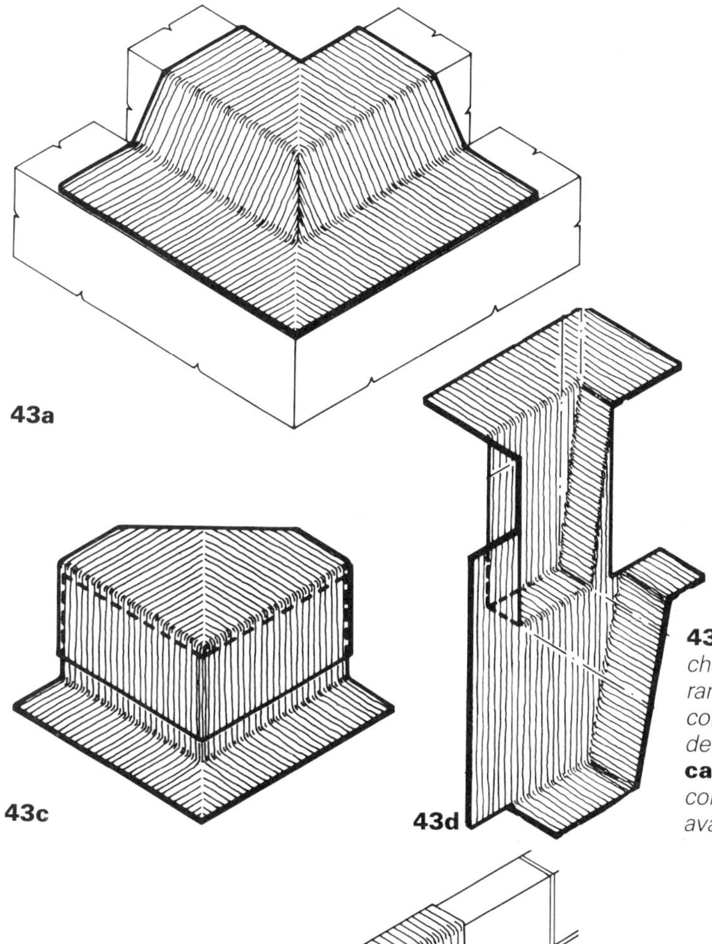

43a

43c

43d

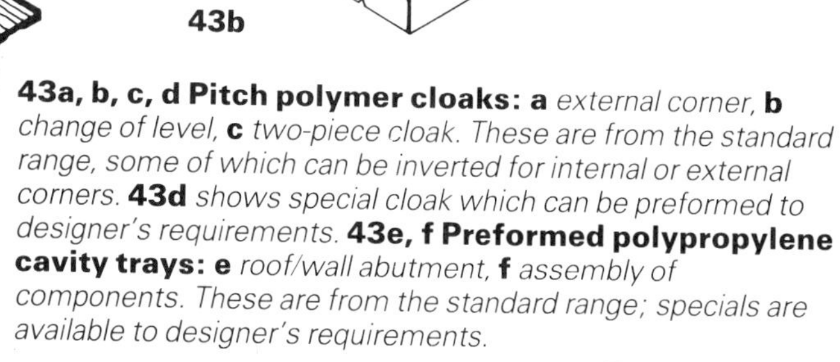

43b

43a, b, c, d Pitch polymer cloaks: a *external corner,* **b** *change of level,* **c** *two-piece cloak. These are from the standard range, some of which can be inverted for internal or external corners.* **43d** *shows special cloak which can be preformed to designer's requirements.* **43e, f Preformed polypropylene cavity trays: e** *roof/wall abutment,* **f** *assembly of components. These are from the standard range; specials are available to designer's requirements.*

43e

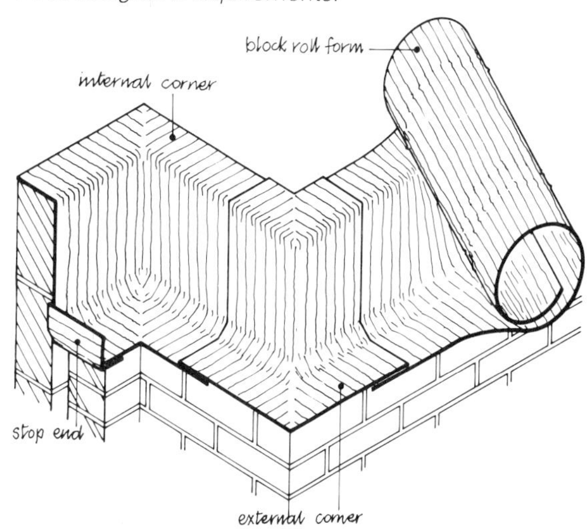

43f

to maintenance problems; and probably in the short term. The introduction of a preformed drip may be a better long term solution. The insistence that any building over three stories must incorporate these devices seems to ignore many, many brick clad concrete framed buildings of up to five storeys which have not had problems although they have not used these elaborate measures. New guidance is being published by manufacturers and trade associations but will have to be very carefully considered by architects and engineers, as there is not yet any clear consensus on how to detail these supporting angles.

4.9 Checklist for dpc detailing

Summarising the advice given in sections 2, 3 and 4, the following checklist for designers is suggested:

1 **Exposure.** Standard of detailing should be related to exposure rating of building (see paragraph 2.2 etc).
2 **Function.** Consider two types of dpcs: those for *rising damp;* and those with the more challenging function of resisting *discharging water,* whether vertically downward, or horizontal (see paragraph 3.3 and 3.4).
3 **Criticality.** Dpcs in critical portions of the building must be considered separately (see paragraph 2.6).
4 **Typical situations.** Draw details for all typical situations in plan and section.
5 **Complex situations.** For complicated corners, junctions, and so on, consider detail very carefully in three dimensions, using isometric sketches.
6 **Sealing.** Remember sealing of laps required in many cases.
7 **Fabrication and installation.** Consider details in relation to formation on site, building in, etc, to ensure practicality of detail.

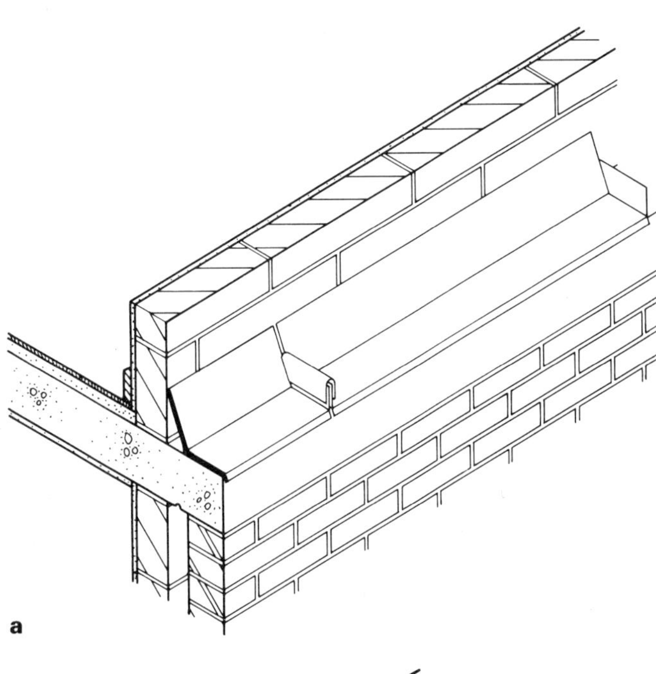

a

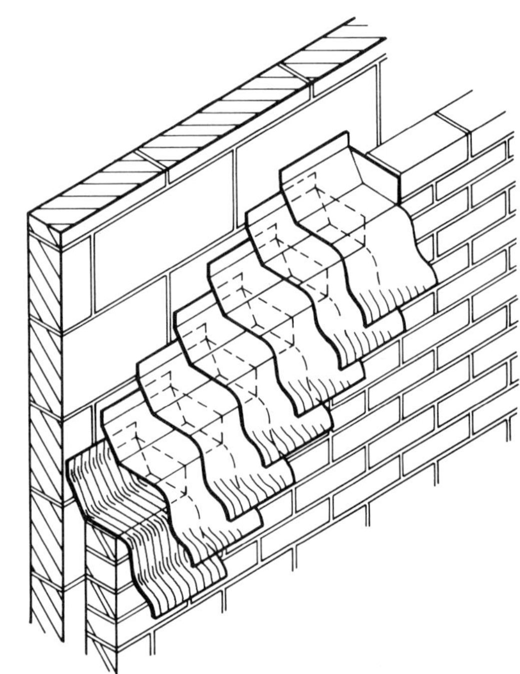

b

44 *Pressed metal trays:* **a** *ring beam, etc,* **b** *roof/wall abutment. Available in galvanised or stainless steel; most forms can have a lead apron welted into steel section to flash over roof tiles, etc. Note the drip over the ring beam joint* **44a,** *and U-shaped clip joints. Sections are 1.0 m long.*

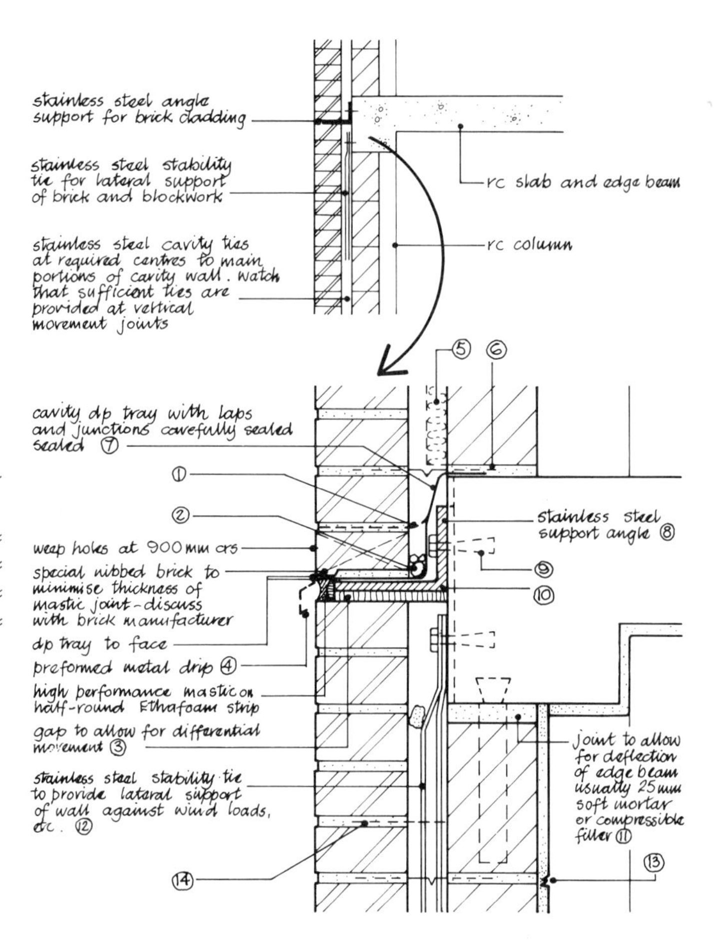

1 For sheltered situations: alternative dp tray level to avoid bolt head. Tray is unsupported and laps difficult to seal. Short lengths of dpc required at lower level to seal junctions of an angle.
2 Mortar droppings very difficult to clean off; probably best to minimise by cavity battens and leave in place the few that do fall into cavity.
3 Gap must be calculated: usually 10–15 mm. Fill with compressible filler.
4 Preformed metal drip will mean less maintenance. Watch for galvanic action between different materials.
5 Possibility of insulation in the cavity. Difficult to install at cavity tray and stainless steel angle.
6 Cavity ties as close to angle support as possible to minimise amount of brickwork not tied back.
7 Cavity dp tray take care bolt head does not damage tray; forms slip plane at angle, hence important brickwork tied back.
8 Size and spacing of fixings depends on amount of brickwork supported. Consider carefully tolerances of fixing angle compared with brickwork below, and face of concrete. Horseshoe shaped shims to height of angle may be necessary.
9 Saw-toothed cast-in slots are available and may give easier adjustment to avoid problems with hitting reinforcing steel while drilling.
10 Angle forms cold bridge. Extending filler and gluing to angle might help to minimise danger of condensation on underside.
11 Some designers do not consider this soft joint to be necessary and prefer to pin up blockwork to underside.
12 Ties come in various types; some are built into inner skin only, with abbey slot in beam soffit. Watch for mortar lodging on any types in cavity.
13 Stability tie may cause cracking at course joints rather than usual crack in internal corner between wall and ceiling. Supports are often specified at each floor level but may only be necessary at every second level.
14 Independent ties to inner and outer skins will probably help to minimise cracks and ensure independent movement.

Special note: Elaborate serrated angles and individual supports at each perpend are being developed. Specifier should inspect actual sites where detail has been used before specifying.

45 *Cavity dp trays to brickwork support angles (see section 4.8).*

5 Specification and measurement

5.1 Specification of dpcs

The specification of dpcs should be much more carefully considered than it normally is (indeed, this is true of specification generally, which is too often delayed to the last minute, when it is left to the quantity surveyor to bring out his 'standard' office specifications for the architect's comments).

Current practice

It is usual to include a fairly brief specification clause in the Brickwork/Blockwork preambles covering:

- type of dpc material and thickness,
- extent of mortar bedding and haunching,
- extent of laps including angles etc,
- termination of dpc (ie recessed, flush or projecting).

The clauses in the National Building Specification 1973[23] gave alternatives for the above items together with alternative clauses for related work, eg:

- keeping cavities clean,
- cleaning bottom of cavity,
- brief descriptions of types of dpcs over openings, at jambs, sills, etc.

It is very important that the specification carefully specify the sealing of dpc laps for downward flow of water (see section 4.6, table VI) although the usual description 'according to manufacturer's instructions' may not be possible as discussed earlier. Clauses should be expanded to warn of the danger of damage to the dpcs when cleaning the cavities. Indeed, more stress should be placed upon the importance of preventing excess mortar droppings reaching the damp-proof course. The specification should state that mortar droppings that do fall on the dpc should be carefully removed before they have time to harden, and contain a clear warning of what action will be taken if this clause has not been followed. The latter is suggested because of the great difficulties which may develop on site if an architect finds on his weekly visit that the clause has not been followed, and that two metres of wall have since been built above dpc. What can he do? Cleaning away the hardened droppings will damage the dpc; but ordering the wall to be pulled down may be politically difficult if clear warning of such a possibility has not been given in good time.

Suggested clauses

The first edition of the *National Building Specification*[19] contained fairly brief clauses, more suitable for simple situations in sheltered sites than for the demanding situations outlined in this series. Firms who have not yet acquired (by subscription, as NBS is now marketed as an ongoing service to subscribers, rather than a simple set of bound volumes) the latest revisions of these original NBS clauses, will therefore not find them satisfactory for the full range of conditions to be met.

The current edition of the NBS section F21 covering dpcs is much improved. However, in view of the new approach to dpc detailing suggested here, with a higher standard of weatherproofing for relatively exposed conditions than that which has usually been demanded in the past, the author suggests on the next page some alterations and additions to the NBS clauses contained in National Building Specification section F21: *Brick/block walling.* These new clauses, it is argued, are more appropriate for buildings classed as standards B or C in **table II.** They are based on the author's own experience, and on various official and unofficial specifications, such as those of the GLC and PSA. The courtesy of the latter, in allowing their documents to be examined and drawn upon, is hereby acknowledged; though neither the GLC or PSA should be held responsible for the technical content of the suggested clauses.

Commodities

The specification clauses given in the NBS under *Commodities* are generally satisfactory but for the reasons discussed under the properties sections (3.3 and 3.4), the author believes it is wise in most cases to specify one manufacturer's material.

The author would suggest that if clay bricks are to be used as a dpc they should be so identified in this section, eg 'clay dpc bricks to BS 3921, table 6, class dpc'. If in situ coatings (eg pitch epoxy/fibreglass) are to be used in place of the more usual materials, they will in the NBS system be included elsewhere. It may be wise to bring them into the brickwork section to which they are most closely related. If weep tubes are to be used they should be specified in the Accessories section, F 21: Likewise preformed 'Cloaks' (ie corners, junctions, etc) should be included under Accessories if they are required. (Note: the specification clauses for dpcs in the NBS Small Works version are virtually identical to those in the full NBS specification).

5.2 Measurement of dpcs

The measurement of dpcs in bills as currently practiced by quantity surveyors following *The standard method of measurement* needs reconsideration. While horizontal dpcs are usually given in linear quantities of a given width, which gives a reasonable picture of the actual job involved, cavity dp trays and vertical dpcs are measured by total area which gives no idea of the work involved. It would be more realistic if cavity dp trays were measured linearly with notes on width, basic shape, etc, and this might be even more appropriate under the latest SMM edition, with sketches possibly included in the bills.

Angles, corners and junctions in cavity dp trays are not normally required to be separately measured, although it would be reasonable for more detailed information to be given where it is known. This would provide the opportunity for dpc work to be properly assessed at tender stage and avoid the problems which can now arise during the construction period. The costs given in **table IV** are based upon actual tender figures and do not, in the author's opinion, reflect the true onsite costs of dpc work.

It is interesting, incidentally, to note that the usual description of a cavity dp tray in bills of quantities is 'cavity gutters' and this may be a more accurate description of their function in severely exposed buildings.

Suggested clauses to be used with the National Building Specification

WORKMANSHIP

F21:5 CAVITY WORK

5051- ALTERNATIVE CLAUSES*

"GENERALLY: Keep cavity and ties free from mortar and debris by lifting battens, or other means approved by the architect".

"CLEANING CAVITIES: Cavities are to be carefully cleaned before mortar sets onto the damp-proof course. Cavities are to be cleaned from top of cavity to holes left in outer/inner skin as required". (If loadbearing walls, check with structural engineer before finalising the specification).

"PROTECTION OF DAMP-PROOF COURSES: Precautions are to be taken to prevent damage to cavity dp trays during cleaning operation. Sharp implements are not to be used for cleaning off mortar droppings. Inspect damp-proof course for damage as work proceeds".

5251, 5252- ALTERNATIVE CLAUSES*

"WEEPHOLES: Form weepholes in perpend joint 75 mm high at the base of the cavity and over external openings, ensuring they are not blocked with mortar".

"WEEPHOLES: Form weepholes in bottom joint at the perpend joint, 10 mm in diameter, directly on top of the damp-proof course, ensuring they are not blocked with mortar".

"WEEPTUBES: Form holes 10 mm in diameter and insert weep tubes ensuring that they are bedded tight onto the damp-proof course and clear of obstructions".

"SPACING OF WEEPHOLES/TUBES: To be at approximately 900 mm intervals beginning approximately 450 mm from ends and other junctions. Weepholes to be aligned vertically above each other in multi-storey situations".

F21:6 DAMP-PROOF COURSES

6201- ALTERNATIVE CLAUSES*

"BEDDING NORMALLY: Fill frogs and/or perforations and flush up brickwork (or rough concrete) to an even bed, using mortar as in brickwork below to receive damp-proof course".

"BEDDING: PARAPETS AND LIGHTLY LOADED WALLS: Fill frogs and/or perforations and flush up brickwork (or rough concrete) to an even bed, using mortar as brickwork below. Lay damp-proof course onto wet mortar and immediately lay first course of brickwork above so that damp-proof course is well bedded between wet mortar".

"BEDDING: ONTO SMOOTH CONCRETE: Flush up any imperfections in concrete to receive damp-proof course".

6152- ALTERNATIVE CLAUSES*

"LAPPING: HORIZONTAL AND STEPPED DPCs: Lay damp proof course in continuous strip with 100 mm laps in length and full laps at angles and complete joint to normal thickness".

"LAPPING: CAVITY DP TRAYS, UNDER COPINGS: Seal 100 mm laps, angles, corners, junctions to prevent water seepage as recommended by manufacturer".**

"LAPPING: VERTICAL DPCs: Are to be in one piece unless this is not possible, in which case lap upper section 100 mm over lower".

6251, 6252, 6253- ALTERNATIVE CLAUSES*

"FACING WORK: TERMINATION OF DPC: Keep leading edge of damp-proof course 10 mm back from face of brickwork".

"FACING WORK: TERMINATION OF DPC: Keep leading edge of damp-proof course flush with face of brickwork".

"FACING WORK: TERMINATION OF DPC: Dress down edge of malleable metal damp-proof course to from drip and mask bed joint".

6301, 6351, 6401, 6451- ALTERNATIVE CLAUSES*

"OPENINGS: AT HEAD: Provide cavity dp tray in one piece extending 100 mm beyond ends of lintels and sloping to weepholes ensuring it is brought well forward at ends ".

"OPENINGS: AT JAMBS: Provide vertical damp-proof course in one piece to jambs of openings in cavity walls, fully lapped behind cavity dp tray at head and over horizontal damp-proof course at sill".

"OPENINGS: TIMBER SILLS: Provide horizontal dpc under sill lapped with self-adhesive bitumen sheet to seal end grain".

"OPENINGS: JOINTED SILLS: Provide horizontal damp-proof course under sills turned up at back and ends".

ADDITIONAL CLAUSES*

"CAVITY DP TRAYS IN PARAPETS, ETC: Where damp-proof course is unsupported it is to be fixed into joint of upper level, drawn across the cavity in a 'Z' shape and continued through outer skin 75 mm lower, ensuring all laps sealed and mortar droppings kept free of cavity".

"CAVITY DP TRAYS AT FLOOR LEVELS, ETC: Provide damp-proof course fixed into joint of inner skin, turned down inner skin and across base of cavity in an 'L' shape continued through outer skin 150/225 mm lower, ensuring all laps are sealed and cavity dp tray well supported by backing construction".

"CORNERS, JUNCTIONS, ETC: In cavity dp trays are to be formed/pre-formed to maintain continuity of damp-proof course with all laps sealed".

* Non-relevant clauses may be omitted

** See table VI an alternative.

6 Installation of dpcs

6.1 General comments

All dpc materials need reasonable care in handling and installation. It must be emphasised that dpcs are a very important part of the building fabric and they need to be carefully installed with good workmanship by the craftsman and careful supervision by the contractor's supervisory personnel. As has been reported by the BRS Advisory Service, a significant number of cases of damp penetration have been caused by faulty or damaged dpcs. The architect should emphasise the importance of the dpc installation at the time work on dpcs begins on site. The bricklayers on all but the simplest sites should be briefed on the dpc installation and mock-ups of corners, junctions, etc, may be required, **36.** All those concerned should then periodically check all aspects of the dpc work.

6.2 Site installation in relation to different dpc materials

The relative properties of the different dpc materials obviously have an effect on their site installation. Some of these aspects have been discussed in relation to the selection of suitable dpc materials in section 3 but certain aspects need to be investigated further.

Weight

It is not often realised that some commonly specified dpc materials are fairly difficult to handle due to the simple fact that they are heavy and cumbersome. A three-metre strip of 350 mm wide bitumen/asbestos/lead dpc (BS 743, type F) suitable for a cavity dp tray weighs approximately 4 kg and can be quite unwieldy on high scaffolding and in awkward situations (some of the most popular 'quality' lead-cored dpcs are even heavier). In contrast, a polythene dpc to BS 743 would weigh only 1/10th as much. This is one of the main reasons for the popularity of polythene dpc with many contractors for simple dpc work in speculative housing, (the other is cost).

Workability

Many site problems regarding the installation of dpcs are related to various aspects of the workability of the dpc materials. As outlined in **table III** workability concerns:

1 the ability of the material to be formed into shape (at high and low temperatures);
2 the ability to keep that shape without restraint; and
3 the ease in forming corners and junctions.

These aspects are not very important for simple horizontal dpcs, **46,** but for cavity dp trays they are critical.

Regarding 1, *malleable metals* can be accurately formed into complicated shapes but this work is time consuming and usually done by a skilled plumber and consequently very expensive. For practical reasons, malleable metals are usually pre-formed to shape by the manufacturer/fabricator or in a site shop. Before building-in, most malleable metals should be primed with bitumen paint to prevent corrosion from freshly laid mortar which is yet another expensive operation. *Bitumen-based* dpc materials are quicker to form than malleable metals and because of their nature they are formed in situ by the bricklayer. They require extreme care when handled in cold weather. The rolls of dpc material must be warmed before unrolling. If cracking is to be avoided, they must be adequately softened with a blow lamp even in warm weather when they are formed into cavity dp trays, etc. *Polythene* dpc material is somewhat difficult to form into shape, because of its resilience but 'creasing' along the line of changes in profile can help. It does not crack easily even at low temperature. *Pitch* and *bitumen polymer* dpc materials, **46,** are fairly resilient but are easier to form into accurate profiles than other flexible dpc materials. Like polythene, they do not crack easily at low temperatures. The newest dpc material, polypropylene, comes pre-creased for cavity tray use.

Regarding 2 above, once the cavity dp tray or other flashing is formed into shape, it is important that it must maintain that shape. *Malleable metals* are obviously very good in this respect. *Bitumen-based* dpcs if adequately softened and bedded to their backing will also keep their shape, although it is difficult to maintain the requisite degree of workmanship to achieve this on site, **47a, b.** The *polymer-based* materials because of their resilience, must be positively restrained to make them keep their proper shape. This is often doen by nailing the dpc tuck-in to the inner skin blockwork and weighing down the sole of the dp tray with loose bricks although this can cause problems if a wet mortar bed is used, **47c.** Care should be taken to prevent any large puckers forming at the front edge of the dpc, **48.** With polymer-based dp trays it is important that the final construction acts as a positive restraint to prevent them springing out of shape.

Regarding 3 above: the ease with which corners and junctions can be formed on site varies considerably depending upon the material used. *Malleable metals* take a great deal of time and skill. *Bitumen-based* materials (especially the hessian reinforced types with plenty of 'free' bitumen) can be made into good corners and junctions if the material is adequately softened and bedded together but again it is time consuming and some skill is required. *Pitch* and *bitumen polymer* dpc materials can be easily cut and the laps glued with contact adhesive or mastic so that corners and junctions are fairly easy to form (for details of the formation of corners, etc, in situ see section **6.3**). They can also be pre-formed in a site shop using the contact adhesive in ideal conditions in this case. In this way the advantages of pre-formed corners, junctions, etc, can be obtained on even a small job that does not warrant manufactured preformed components. *Polythene,* on the other hand, because of its great resilience is very difficult to form into corners and junctions in situ. The only practical technique is heat welding under controlled conditions though one type of polypropylene dpc is vacuum formed to shape.

46 *Pitch polymer strip dpcs being laid. In case of bitumen-based dpc materials, such rolls must be prewarmed in cold weather to avoid cracking.*

47a, b *Incorrect and correct ways, respectively, of fixing bitumen-based dpc tray. Details shown are for wall on suspended slab, where dpcs are intended to deal only with downward rain flow, not with rising damp; see dotted dpc.*

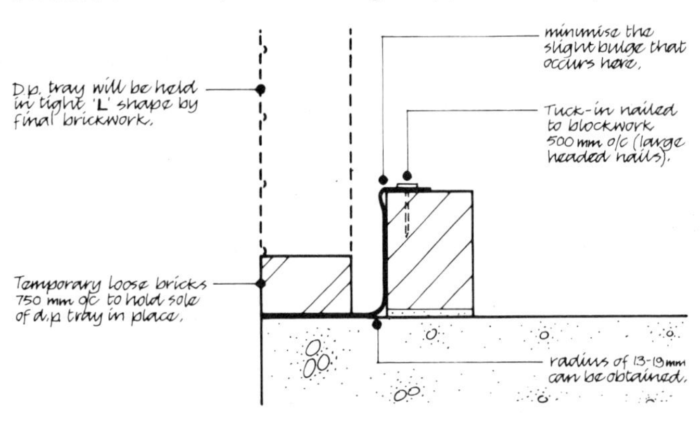

47c *Shows restraining methods required to keep polymer-based dp trays to correct profile.*

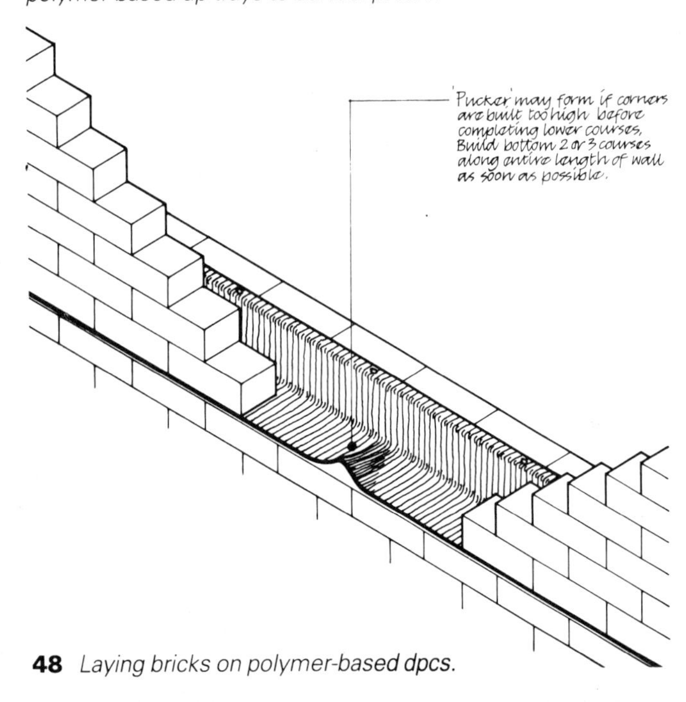

48 *Laying bricks on polymer-based dpcs.*

6.3 Site installation in relation to dpc detailing

As mentioned earlier, dpc detailing should take account of the practical aspects of the installation on site and the previous sections on dpc selection and detailing have stressed this. Many aspects of the detailing of dpcs and related work should be carefully considered by all members of the building team and the following points are highlighted with particular reference to clerks of works and the contractor's supervisory staff. They serve also as a useful guide for spot checks on the architect's site visits.

Dpc installations at ground level

1 Ensure that a good overlap is achieved between the slab dpm and wall dpc. Take care that the exposed membranes are protected from damage before completion of construction.
2 Dpcs in timber frame construction require particularly careful installation as many different materials are involved with many fixings required. Watch for perforations of dpcs, **49.**
3 Ensure good dpc installations to freestanding walls including sandwiching dpc between wet mortar, sealing of laps to dpcs under copings and sufficient laps with vertical dpc if the wall retains some earth, **15.**

Dpc installations to windows

1 For cavity dp trays to lintels and sills use a single length of dpc material so no laps are necessary.
2 If laps necessary in vertical dpcs, provide a 100 mm overlap. If brickwork ties necessary at jamb vertical dpc, use cranked ties with overlap, **50.**
3 Check junctions and overlap at head (and at sill if dpc provided there), **52a.**
4 Check that profile of cavity dp tray over lintel at end discharges towards outer skin, **52b.** (The 'L'-shaped profile recommended elsewhere is *not* to be used here.)
5 If windows to be installed after opening is formed, great care is required in placing the window to ensure dpcs are not damaged and end up correctly positioned, **51.**

Dpc installations to parapets

1 During construction check that all laps are carefully sealed, both directly under coping and in any cavity dp trays. In severely exposed situations, testing of dp tray laps may be advisable, **53.**
2 Ensure that dpcs are sandwiched between wet mortar, particularly under coping.
3 Ensure that coping dpc is supported over cavity.
4 Carefully inspect various components of roofing kerb/wall dpc junction. Protection may be necessary at certain stages in construction, **54.**

Laps in dpcs

1 Check adequate overlapping for dpcs for rising damp.
2 Ensure bricklayers understand method of sealing laps for downward flow of water. Figures **55a** and **b** show methods used for bitumen-based and pitch polymer dpc materials respectively.
3 Check sealing of laps frequently in cavity dp trays. On severely exposed sites, contractor should make spot tests by filling dp tray with water, **53.**

Bedding of dpcs

1 If dpcs are to be laid on a wet mortar bed, ensure they are bedded in as level as possible and are not disturbed when first course of bricks are laid onto it.
2 If dpcs are to be laid on bricks with perforations or frogs these must be evenly flushed up, **56.**
3 If dpcs are to be laid on dry mortar bed or concrete; ensure bed is smooth and free of indentations, pockets, projections, etc, **57.**

Method of forming corner and junctions on site

1 Figure **58** shows how an external corner is usually formed on site in flexible dpc materials. A separate 'L'-shaped gusset and a small patch are added to the dp tray proper to ensure a continuous membrane. When making 90 degree cuts in dpc material a rounded cut should be made at the internal angle, to help

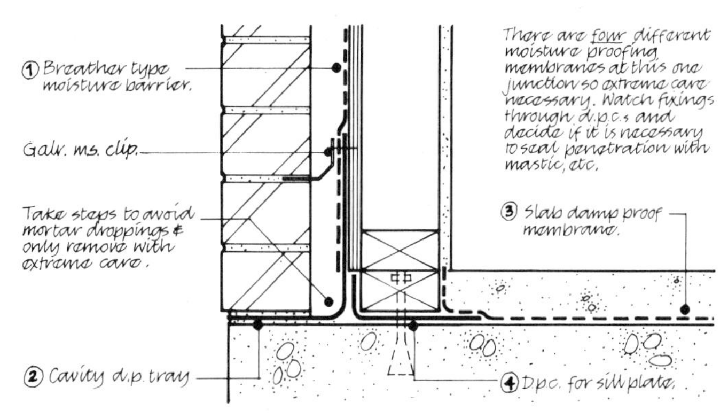

49 *Dpcs at base of brick-clad timber frame. It is important to watch for perforations of the dpc, and if necessary to seal them with mastic.*

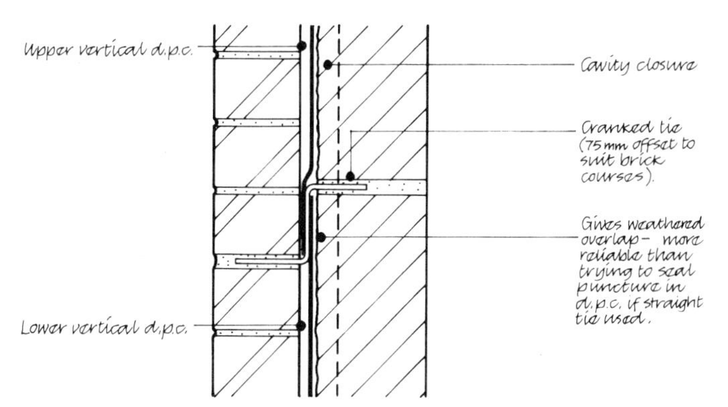

50 *Use of cranked ties at vertical dpc allows weathered overlap with no need to puncture dpc.*

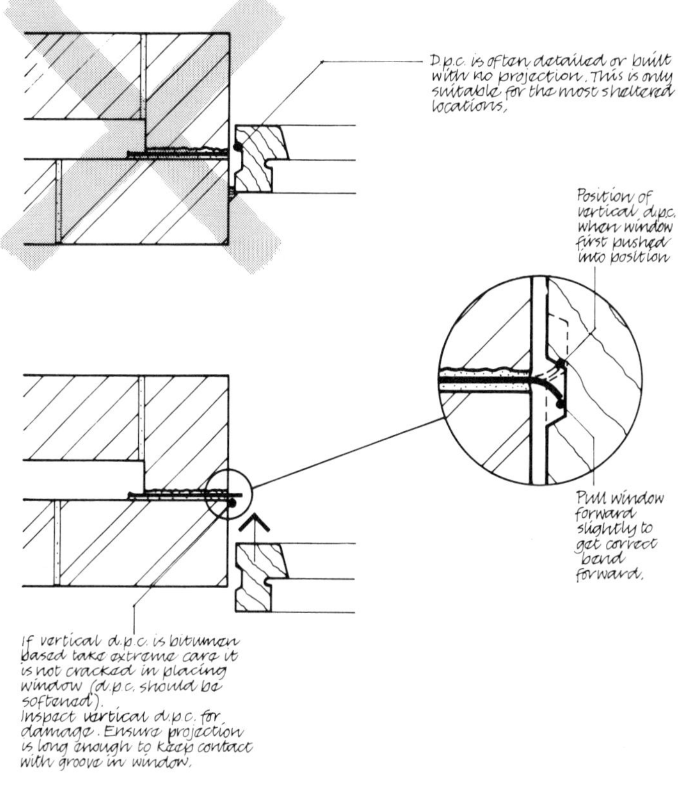

51 *Installation of windows into prepared openings. Detail at top is suitable only for sheltered locations. Detail above is safer, but requires great care while inserting window. For windows built in during course of wall construction, see earlier details* **19a, b.**

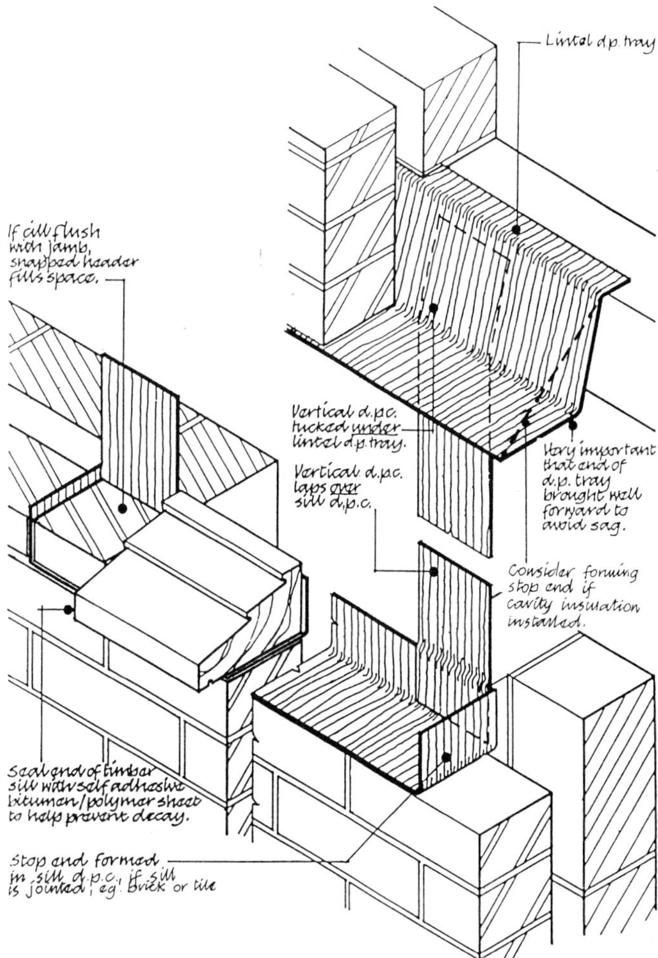

52a *Correct junction of dpcs at window head and sill, showing overlapping of various pieces to form watertight assembly. Specific profiles and arrangement will of course depend on lintel type, and length of sill projection to either side of window opening (in above detail, sill extends one half-brick to either side of opening). End of timber sill must be waterproofed.*

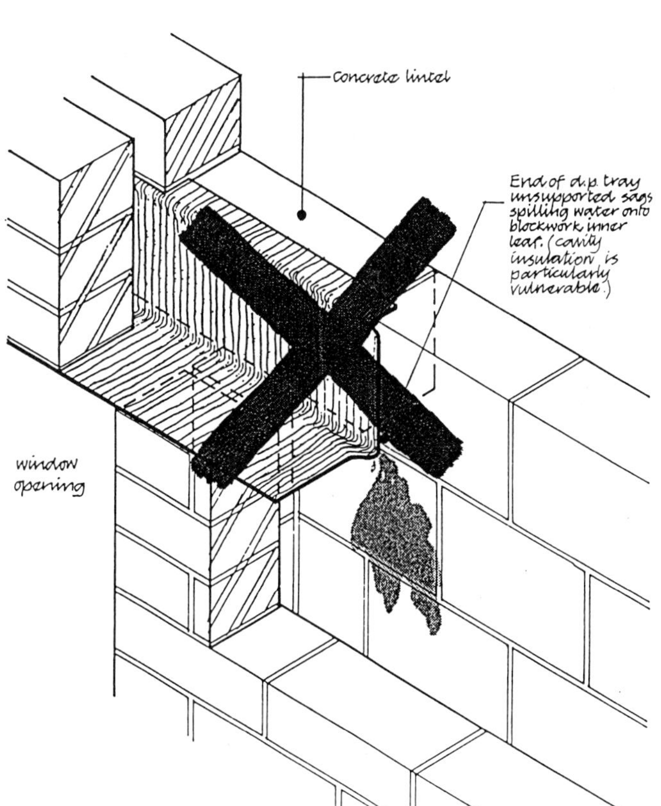

52b *Possible damp at window head if tray at lintel is not pulled well forward, and therefore allowed to sag at end.*

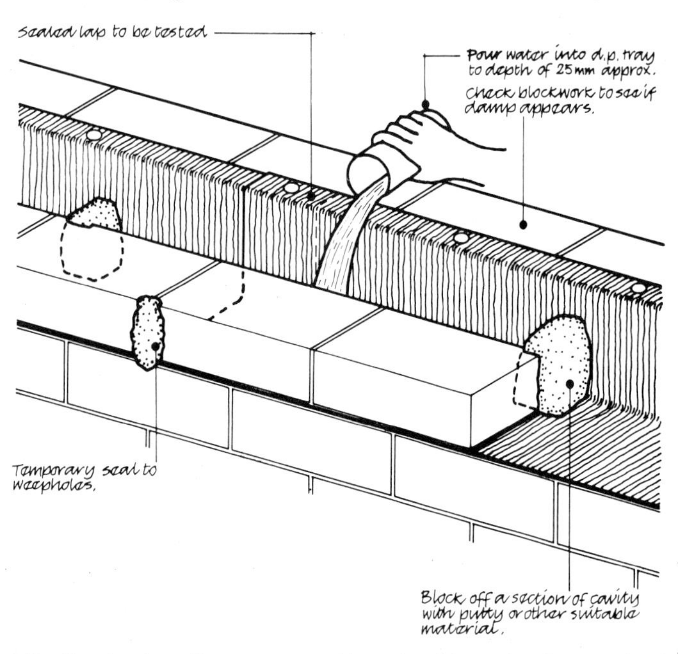

53 *Testing laps (or corners and junctions) in cavity dp trays for leakage. This would only be done in severely exposed locations.*

Ensure roofing is well worked into chase.
Ensure sound pointing on completion of all work.
Check cover flashing is well fixed so it won't work loose.
NOTE:
If cover flashing has not been built into work, extreme care is necessary to insert it. In this case, inspect edge of d.p.c. for damage & carefully wedge flashing under it, then point joint with well packed mortar with an "expanding agent". (not recommended).

54 *Inspection of roof-kerb/dpc junction during construction.*

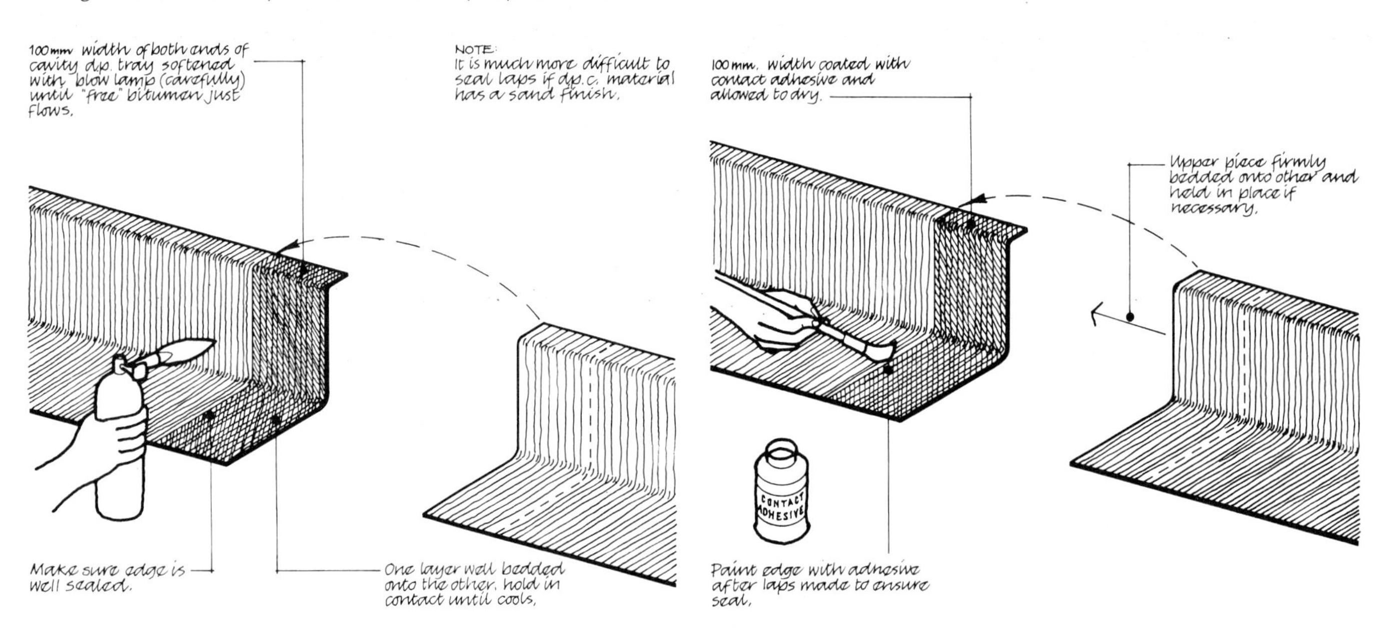

55a, b *Sealing of laps in, respectively, bitumen-based dpcs (using blow lamp); and pitch-polymer dpcs (using adhesive).*

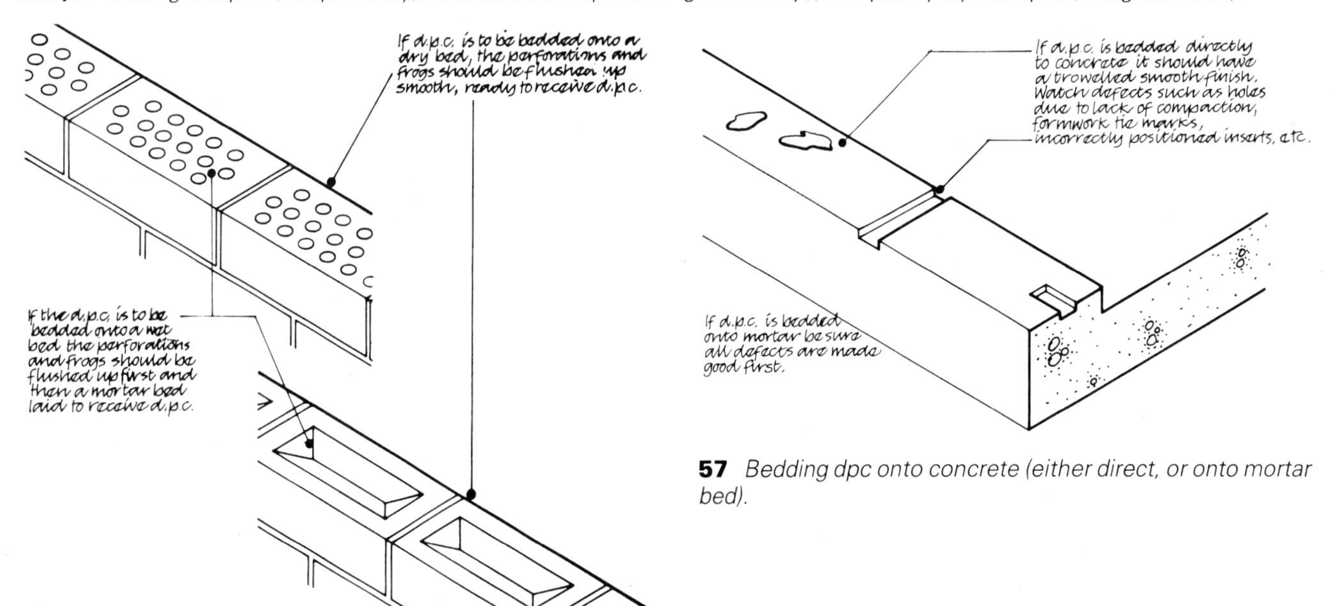

56 *Bedding dpc onto brickwork (either onto a dry bed, or a wet bed).*

57 *Bedding dpc onto concrete (either direct, or onto mortar bed).*

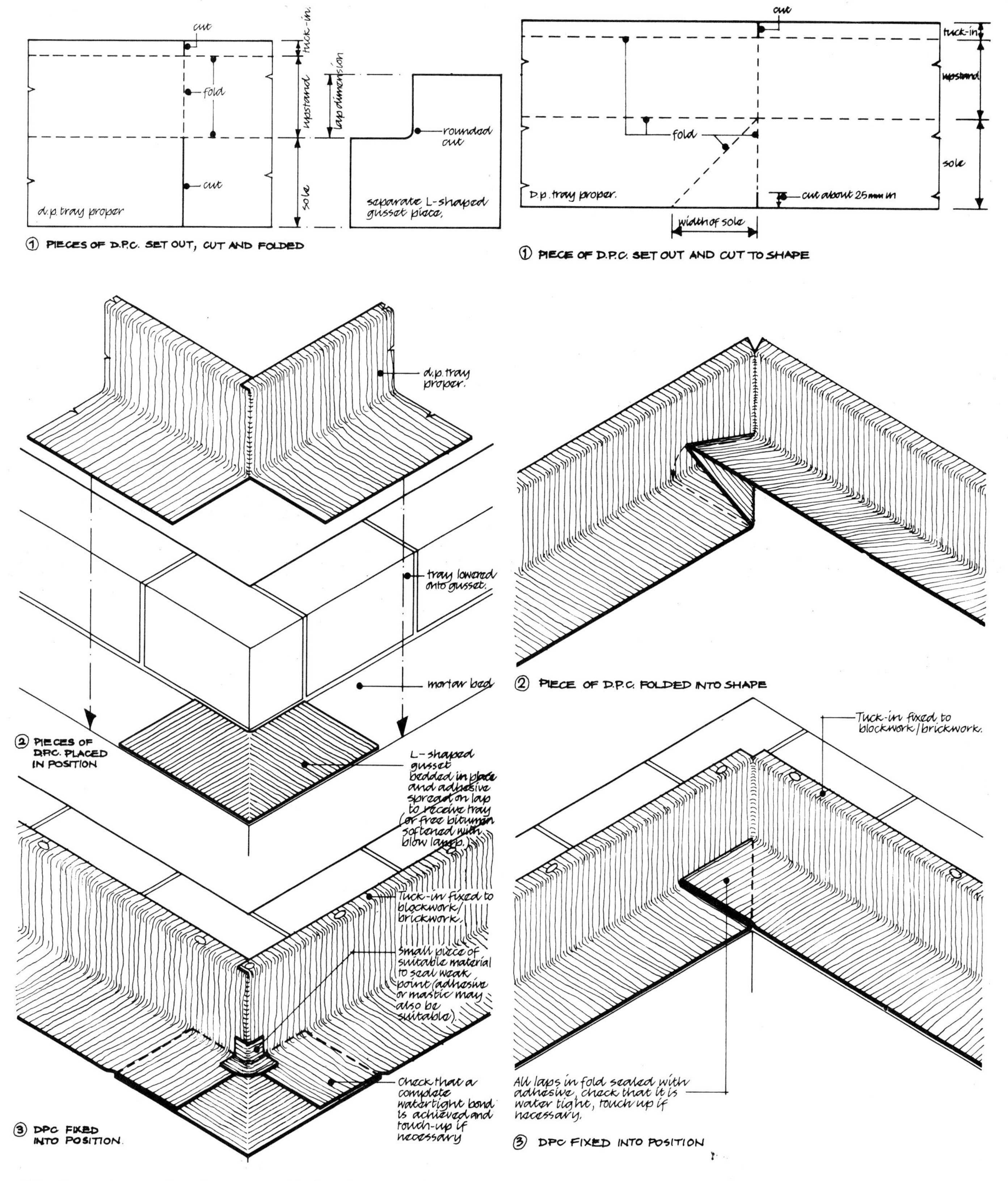

58 *Formation on site of an external (salient) corner, shown in three successive stages.*

59 *Formation on site of an internal (re-entrant) corner, shown in three stages of operation.*

prevent accidental tears. It can be seen that if an 'L'-shaped profile is used, the geometry of forming corners is not difficult but with the standard 'Z' shape, the geometry is much more complex and difficult to achieve on site.

2 Figure **59** shows how an internal corner is formed on site. In this case a separate gusset piece is not required (with very thick bitumen-based dpc materials the junction may be cut rather than folded and in this case it is important that the point of coincidence of the cuts at the back of the cavity is effectively sealed with adhesive, mastic or patch).

3 Figure **60** shows how a stop end at a column or other obstruction can be formed.

4 Other junctions, changes in level, flashings for pre-cast claddings, etc, can be made in situ using the principles covered in the three cases above. Care and time *must* be taken to form all these details. On large sites, it may be desirable to allocate the work to 'specialists' to help ensure consistent, watertight details. It is also possible as mentioned earlier to pre-form the details as above in a site shop using jigs, etc, so that better working conditions are achieved.

5 If preformed junctions are used on site (either the manufactured ones or those from a site workshop) it is very important that effective laps are made where it joins the normal run of dp tray, **61**.

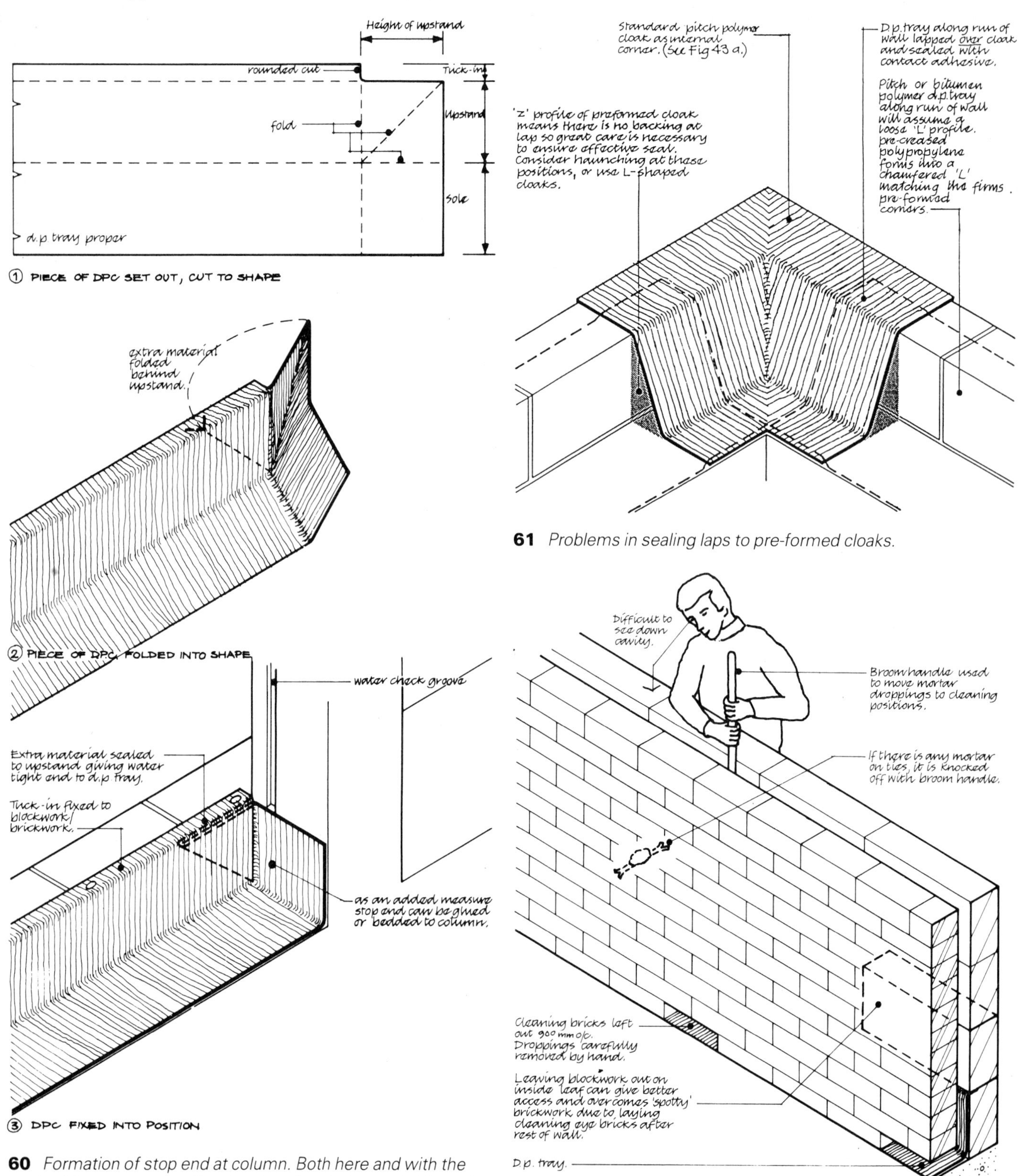

60 *Formation of stop end at column. Both here and with the earlier* **39c,** *water run-off at the chamfered stop-end at foot of vertical groove, onto the dpc, should be considered. In sheltered locations, gluing the dpc upstand to the column face should be sufficient. Otherwise, a horizontal groove may be formed in the concrete column, and the top edge of the dpc upstand tucked in and fixed along its full width.*

61 *Problems in sealing laps to pre-formed cloaks.*

62 *Cleaning cavities of mortar droppings.*

Mortar droppings fall on sand and do not adhere to d.p. tray. Sand acts as a cushion during cleaning out (but how to make certain this is done?).

Mortar droppings can be broken with broom handle without puncturing d.p.c material

50 mm sand placed in tray as wall started.

63 *Protection of dp tray with sand bed, for cleaning out.*

6.4 Damage to dpcs

It is important that all levels of the contractor's staff on site realise the significance of careful handling and protection of dpcs during construction.

Different dpc materials have to be considered differently in relation to the protection they are given, eg in cold weather an exposed edge of a bitumen-based vertical dpc could easily crack with the slightest accidental knock while a polymer-based vertical dpc would be much more resistant to cracking. Hessian reinforced bitumen dpcs are much more resistant to tears and

Table VII: Checklists for site installation

		Contractor	Clerk of works	Architect
Ordering	**1** Order specified dpc materials including adhesives, preformed corners etc	•		
	2 Check that specified dpc materials are received on site and adequately stored	•	• *occasionally*	
Planning	**3** Stress the need for good workmanship in dpc installation at site meeting			•
	4 Decide on hierarchy of supervision of dpc installation	•		
	5 Foreman briefs bricklayers and others involved on dpc installation generally	•		
	6 Mock-ups of formation of laps, corners, etc are made and discussed[1]	•	•	•
	7 Decide on method for keeping cavities clean, and/or cleaning cavities afterwards	•		
	Frequency of checking for **8** to **14**	*continuously*	*frequently*	*occasionally*
Actual installation	**8** Check that specified materials are actually used	•	•	•
	9 Check formation of basic dpcs including: correct widths, lapping, bedding/haunching	•	•	•
	10 Special checks on more critical dpc details including: parapets, cavity trays, corners/junctions	•	•	•
	11 Check that cavities are being kept as clean as specified	•	•	•
	12 Check method of cleaning cavities, watching for possible causes of damage; check that mortar droppings are cleared before they harden	•	•	•
	13 Inspect for damage to dpcs (especially cavity dp trays, exposed vdpcs, flashings[2]	•	•	•
	14 Check weepholes for spacing, being clear of mortar, etc	•	•	•

(1) *Not required on sheltered sites and usually only where there is a high degree of repetition.*
(2) *Testing may be necessary on exposed sites or on other sites where damage suspected.*

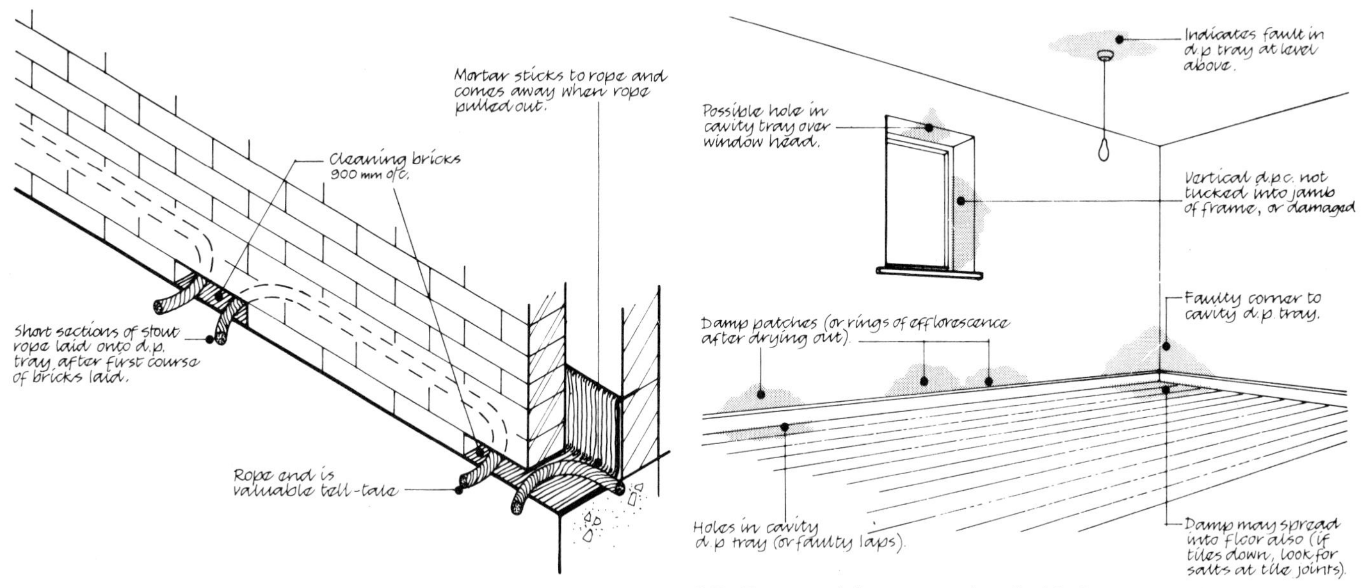

64 *Alternative method to* **63**: *protection of dp tray with rope.*

65 *Pattern of damp associated with damaged or faulty dpcs.*

punctures than fibre or asbestos reinforced ones. In other words, the contractor's staff must be as aware of the physical properties of dpc materials as the designer.

It cannot be over emphasised that the cleaning of mortar droppings off the cavity dp tray is the most dangerous period for damaging dpcs. This usually occurs at window heads, parapets and solid floors. Small amounts of mortar droppings on the dpc at the base of the cavity are of no consequence and could be left, but it is now commonplace to see 150 mm or more of mortar deposited on the dp tray.

This occurrence is regrettable and contractors should either insist that lifting battens are used or devise other means of keeping the cavity clean or mortar droppings off the dp tray. Careful hosing of the cavity before the mortar has set has been tried but this introduces even more water into the construction process. It is important that the droppings are carefully removed before the mortar dries hard, otherwise they will adhere to the brick/block faces and the dpc.

Attempts to remove them towards the cleaning positions will then result in torn or punctured dpcs. The 'tool' used to remove the mortar droppings should have a rounded end, eg a broom handle, and on no account should sharp battens, reinforcing rods, etc, be used, **62.**

Alternatively, the base of the cavity can be temporarily filled with sand or newspapers or rope laid onto the dp tray to keep the droppings from adhering to the dpc, **63** and **64.** The rope method has the advantage of visual evidence that the dpc is being protected and when it has been removed that the base of the cavity has been cleared. The layer of sand without rope can be overlooked, and left in place *with* the mortar droppings. Cavity dp trays should be inspected for damage as work proceeds although because of the confined spaces this is difficult. If damage is suspected (for example an operative is observed cleaning the cavity with a sharp implement) then the suspect cavity should be tested by filling with water and subsequently checking for dampness on the inner face of the cavity wall.

6.5 A note on damp penetration

Damage to dpcs and other bad workmanship in relation to dpcs is often not evident until after the buildings are virtually finished with weather-proofing and finishes completed and scaffolding removed. Then the inner walls are found to be damp after long periods of wind-driven rain. The pattern of damp is usually a series of crescents caused by the water soaking up the inner skin from the points of damage or faulty workmanship, **65.** In some cases, of course, the problem may be faulty detailing. If decorations are completed the damp may not be visually evident until drying out occurs and efflorescent salts are forced to the surface. Damp meters can be used as an aid to tracing damaged or faulty dpcs but the readings must be carefully considered as they can be affected by other factors to give misleading readings. Water penetration due to faulty dpcs can also be confused with condensation problems.

If damp penetration occurs, dpc repairs are expensive and it is sometimes tempting to consider the use of water repellents, eg silicones, to try to seal the face of the walls. In areas of moderate and severe exposure this treatment is unlikely to be successful and a programme of cutting out the brickwork to expose and repair the dpc is the only real possibility.

6.6 Checklists for site installation and inspection

Table VII is a brief checklist for the use of all concerned with the site installation of dpcs.

7 Directions for the future

It must be obvious that more knowledge is needed about the properties of dpc materials currently available and this knowledge must be presented in a form that makes intelligent comparisons possible. More attention must be given to the selection, detailing, specification and inspection on site of dpcs than has in many cases been true in the past. The true function of the dpc and related work should be observed by site monitoring of typical situations as experience shows that laboratory testing can be misleading. On building sites the contractor should ensure that there is adequate supervision to maintain a high quality of workmanship on this vital part of a building. Standards of workmanship should be improved. Continued development of new materials for dpcs is also desirable as no material currently on the market is without some disadvantages. It seems to the author that three possible courses are open:

- further types of plastic/rubber sheeting with superior physical properties to existing flexible dpcs, particularly workability and toughness;
- self-adhesive strips developed from existing metal foil/bitumen or polythene/bitumen sheet which will make laps, corners, junctions easier to achieve on site;
- multiple spray coatings which will provide a dependable jointless reinforced coating in one operation.

Since the first edition of this book there have been several developments. For difficult situations such as roof to wall junctions, more preformed cloaks are now available. One new material, polypropylene, has been introduced. However, much work remains to be done if more dependable and waterproof buildings are to be commonplace.

8 Product lists

Table VIII: List of manufacturers of flexible dpcs

Anaplast Ltd, 14 Kyle Road, Industrial Estate, Irvine KA12 8NJ (0294 72231)

D. Anderson & Son Ltd, Barton Dock Road, Stretford, Manchester M32 0YL (061 865 4444)

Brett Martin Ltd, Higgins Lane, Burscough, Lancashire L40 8JB (0704 895345)

British Cellophane Ltd, Plastics Film Division, Bath Road, Bridgwater, Somerset TA6 4BA (0278 424 321)

British Hydroflex, Appley Lane North, Appley Bridge, Lancashire WN6 9AB (02575 2333)

Callenders Bituminous Products Ltd, 455 Wick Lane, London E3 2TF (01 980 5548)

Carter Bros Billingshurst Ltd, Wisborough Green, Billingshurst, West Sussex RH14 0AY (0403 700 551)

Cavity Trays Ltd, Administration Centre, Vale Road, Yeovil, Somerset BA2 5HU (0935 4769)

Colas Products Ltd, Galvin Road, Slough SL1 4DL (0753 71711)

DMS Building Components Ltd, Clinton Road, Leominster, Herefordshire HR6 2RT (0568 4051)

ICI Plastics Division, Visqueen Products, Six Hills Way, Stevenage, Hertfordshire SG1 2DB (0438 3400)

Marley Waterproofing Products, P.O. Box 17, Otford, Sevenoaks, Kent TN14 5EW (0732 451033)

Permanite Ltd, Mead Lane, Hertford, Hertfordshire SG13 7AU (0992 50511)

Ruberoid Building Products Ltd, Brimsdown, Enfield, Middlesex EN3 7PP (01 805 3434)

Sitco Building Materials Ltd, The Plantation, Curdridge, Southampton SO3 2DT (04892 6117)

Timloc Building Products Ltd, Rawcliffe Road, Goole, North Humberside DN14 6UQ (0405 5567)

Vulcanite Ltd, Trident Works, Seven Stars Bridge, Wigan, Lancashire WN3 5AF (0942 462292)

Zedcor Marketing Co Ltd, 73a High Street, Witney, Oxfordshire OX8 6LR (0993 71471)

Table IX: Classified list of flexible dpc materials and preformed dpcs

Bitumen based (to BS 743)	**Manufacturer**	**Product**
Type A: Bitumen/hessian	D Anderson	Basite
	British Hydroflex	Hessian based
	Callender	Callendrite dpc
	Callender	Nubit
	Colas	Hydrotite hessian
	Permanite	Permaseal
	Ruberoid	Pluvex No 1
	Vulcanite	Bituna
Type B: Bitumen/fibre	D Anderson	Anderite
	British Hydroflex	Fibre based
	Callender	Stormax
	Colas	Hydrotite fibre
	Permanite	Challenge
	Ruberoid	Pluvex No 2
	Vulcanite	Fibakore
Type C: Bitumen/asbestos	D Anderson	Bestos
	British Hydroflex	Asbestos based
	Callender	Barchester
	Colas	Hydrotite asbestos
	Permanite	Asbex
	Ruberoid	Astos
	Vulcanite	Ascot
Type D: Bitumen/hessian/lead	D Anderson	Ledbit
	Callender	Ledtrinda
	Callender	Ledumite No 1, No 2, No 3[1]
	Permanite	Permalead
	Ruberoid	Pluvex No 3
	Vulcanite	Leadlinde hessian
Type E: Bitumen/fibre/lead	D Anderson	Leadandrite
	Callender	Ledkore A, B, C[1]
	Callender	Libra[1]
	Callender	Nuled
	Permanite	Lead Challenge
	Ruberoid	Pluvex No 4
	Vulcanite	Leadlinde fibre
Type F: Bitumen/asbestos/lead	D Anderson	Ledbestos
	Callender	Trindos
	Permanite	Lead Asbex
	Permanite	Astos Leadlined
	Vulcanite	Ascot Lead
Miscellaneous bitumen based	**Manufacturer**	**Product**
Bitumen/hessian/aluminium	Callender	Alumite
Bitumen/fibre/aluminium	Callender	Alukore
Bitumen/asbestos	Permanite	Permagrip[2]
Polymer based	**Manufacturer**	**Product**
Polythene or polyethylene (most to BS 743)	Anaplast	Brick Grip[2]
	D Anderson	Dri-life
	Brett Martin	Mardamp
	British Cellophane	Duraphane 2000E (was LD20E)[2]
	Carters	Super Rhombos K.M.[3]
	ICI	Visqueen 2000T
	Permanite	Polythene
	Ruberoid	Ruberthene
	Sitco	Sitex[2]
	Vulcanite	Vulcathene
	Zedcor	Zedcorse[2]
Polypropylene	Timloc	Ring beam cavity trays[2]
Pitch polymer	D Anderson	Pitch polymer[2]
	Marley	Aquaguard 125[2]
	Marley	Marleycourse
	Permanite	Permaflex[2]
	Ruberoid	Hyload[2]
	Vulcanite	P.P.D.
Bitumen polymer	Permanite	Permabit[2]
Preformed dpcs, junctions, etc[4]	**Manufacturer**	**Product**
Pitch polymer (flexible)	Ruberoid	Hyload cloaks[2]
Polypropylene (semi-rigid)	Timloc	Preform[2]
Pressed steel (rigid)[5]	Cavity trays	Types E,X,Y,P, etc
Grp (rigid)	D.M..S.	Sillform

This list may not be exhaustive, but all major manufacturers are included. Some firms listed market the product, but do not actually manufacture it. It is wise to compare samples within each classification before specifying.

Notes

1 Material heavier and/or thicker than BS 743 minimum specification.
2 Product covered by an Agrément certificate.
3 Product kite-marked by BSI.
4 Pvc cavity closers are also produced, which can act as a vertical dpc at window or door openings. Many pressed steel lintels have a secondary function as a cavity dp tray at window and door heads.
5 Galvanised or stainless steel is available.

Table X: Summary of Agrément certificates*

Certificate	Product and company	Description	Validity
No 76/421	**Permaflex** Permanite Ltd Lea Road Waltham Abbey Essex For use as a flexible damp-proof course in brick, block, masonry or concrete walls.	Flexible damp-proof course	Valid until further notice
No 80/727	**Permabit 50** Permanite Ltd Lea Road Waltham Abbey Essex EN9 1AY This certificate replaces certificate no 76/389 and relates to a flexible damp-proof course for brick, block, masonry or concrete walls. Installation is in accordance with CP 102: 1973. The product is a black sheet material which remains flexible at low ambient temperatures, does not become tacky in warm ambient conditions, will not exude under load and will withstand considerable movement of the wall. In normal circumstances, it will remain effective during the life of the building.	Damp-proof course for walls	Expires 1/5/83
No 80/728	**Sitex** Sitco Building Materials Ltd The Plantation Curdridge Southampton SO3 2DT Sitex is a polyethylene flexible damp-proof course for use in brick, block, stonework or concrete walls in horizontal, vertical or stepped positions.	Flexible damp-proof course	Expires 1/5/83
No 80/729	**Zedcourse** Zedcor Marketing Ltd 73a High Street Witney Oxfordshire OX8 6LR Zedcourse is a polyethylene flexible damp-proof course for use in brick, block, stonework or concrete walls in horizontal, vertical or stepped positions.	Flexible damp-proof course	Expires 1/5/83
No 80/732	**Timloc Cavity Trays** Timloc Building Products Ltd Rawcliffe Road Goole North Humberside DN14 6UQ Timloc polypropylene cavity trays are for use at abutments between roof and cavity wall and other similar positions. Most are either preformed or precreased to suitable profiles.	Cavity dp trays	Expires 1/6/83
No 80/777	**Permagrip** Permanite Ltd Mead Lane Hertford Hertfordshire SG1 7AV Permagrip is a bitumen/asbestos damp-proof course particularly intended for use in lightly loaded constructions such as piers, garden or parapet walls.	Flexible damp-proof course	Expires 1/11/83
No 81/928	**Anaplast Brickgrip** Anaplast Ltd 14 Kyle Road Industrial Estate Irvine KA12 8NJ Brickgrip is a polyethylene flexible damp-proof course for use in brick, block, stonework or concrete walls in horizontal, vertical or stepped positions.	Flexible damp-proof course	Expires 1/1/85
No 82/940	**Hyload** Ruberoid Building Products Ltd 1 New Oxford Street London WC1A 1PE This certificate replaces certificate no 73/173 and relates to Hyload used as a flexible damp-proof course in brick, block, masonry or concrete walls of both solid and cavity construction in horizontal, vertical or stepped positions including cavity trays. It is installed in the same manner as bitumen sheet damp-proof courses and has the advantage that, during the installation process, it remains flexible at low ambient temperatures. It is tough, not easily damaged and provides an effective barrier to the transmission of liquid water and water vapour.	Flexible damp-proof course	Expires 1/2/85
No 82/1038	**Aquagard** Marley Floors Ltd South Park Sevenoaks Kent This certificate renews certificate no 70/78 and relates to Marley Aquagard 125 flexible damp-proof course material. The product is used in brick, block, masonry and concrete walls generally in accordance with BS Code of Practice 102: 1973, and will prevent the passage of liquid water or water vapour, while remaining flexible at low ambient temperatures.	Damp-proof course: flexible	Valid until further notice
No 82/1047	**Duraphane 2000E (was LD20E)** British Cellophane Ltd Bath Road Bridgwater Somerset This certificate relates to LD20E, a flexible polythene damp-proof course for walls, for use in brick, block, stonework or concrete walls in horizontal, vertical or stepped positions. The product is a black, low density polythene sheet, with both faces embossed with a diamond pattern, which provides a key for wet mortar. The product will withstand considerable movement of the wall, will not extrude under load, and will in normal circumstances remain effective during the lifetime of the building.	Damp-proof course for walls: polythene	Expires 1/2/85
No 82/1048	**Anderson Pitch Polymer DPC** D. Anderson & Son Ltd Barton Dock Road Stretford Manchester M32 0YL A pitch polymer dpc material for use in brick, block, stonework or concrete walls of both solid and cavity construction, in horizontal, vertical or stepped positions, with the exception of cavity trays.	Flexible damp-proof course	Valid until further notice

Note: From 1 May 1982 all new Agrément certificates will be valid 'until further notice', subject to spot checks by the Agrément Board that no changes have occurred in formulation and manufacture. Do not work from these abstracts. For final specification, use the certificates.

Permanite

HIGH PERFORMANCE DAMP-PROOF COURSES

The Permanite range of high performance dpc's has been developed and designed after many years of research to meet the exacting requirements of differing types of constructions. By specifying Permanite dpc's you can be confident that this 'make-or-break' component will give you first-class protection against moisture from above or below ground, in any structure.

Polymer dpc's are tough, clean to handle and easily cut and shaped on site. Their surface finishes provide excellent adhesion to mortar and they will not extrude under heavy loading. Permabit and Permaflex have excellent resistance to thermal movement, a very high degree of elasticity and are virtually indestructible when embedded into the wall.

Permabit
A bitumen polymer dampcourse composed of bitumen, ethylene propylene, synthetic fibres and other additives callendered into a homogenous flexible membrane. Permabit retains its performance and durability under all site conditions, including low temperature flexibility down to as little as 20°C.

Nominal thickness 1.25mm
Nominal weight 1.60 kg/m^2
Roll length 20 metres
Available in all standard wall widths

Permaflex
A pitch polymer dampcourse composed of pitch, PVC, synthetic rubber, fibres and other additives callendered into a homogenous flexible membrane. Permaflex has exceptional qualities, including low temperature flexibility down to –15°C.

Nominal thickness 1.25mm
Nominal weight 1.50 kg/m^2
Roll length 20 metres
Available in all standard wall widths

Permagrip
Permagrip is an asbestos-based dampcourse surfaced with a coarse sand finish designed to give exceptional adhesion to mortar. Permagrip has excellent resistance both to slip along the length of the wall and to tensile stress across the width of the dampcourse.

Nominal thickness 2.15mm
Nominal weight 2.20 kg/m^2
Roll length 8 metres
Available in all standard wall widths

For full technical information on Permanite High Performance and Standard DPC's please contact:
Permanite Limited, Mead Lane, Hertford, Herts SG13 7AU
or telephone Hertford (0992) 50511.

Dampness in buildings

In this study, FRED LAWSON, formerly lecturer in environmental studies, University of Surrey, discusses all forms of dampness in existing buildings: rising damp, penetrating damp, roof failures, plumbing failures and condensation. These are sometimes related to or confused with dampness due to defective dpcs.

Causes of dampness are described in each and remedies are recommended.

9.1 Introduction

With improvements in the standard of living, the demand for better accommodation and decoration has brought a greater awareness of building defects. Improved techniques of building construction have almost eliminated some of the more obvious modes of entry of water from outside, for example rising ground water and penetrating rain, but failure of these measures sometimes occurs and usually has serious results.

Other forms of dampness from internal sources, such as the internal atmosphere, now create widespread concern. Condensation is the most common source of dampness in modern buildings and is one of the main causes of complaint in postwar housing.

Apart from modern buildings, it is important to bear in mind the considerable demand for modernisation and improvement of property particularly the so-called 'twilight' areas of our towns and cities which are now being scheduled for redevelopment. Damp-proofing techniques are particularly important here.

A good deal of literature has been written on the theory of dampness and the object of this analysis is to summarise the main features by which dampness can be identified and the practical remedies which can be applied in each case.

9.2 Rising damp

Absence or failure of damp-proof course in walls. This is shown by a continuous damp area extending upwards from the base of the walls. Both internal and external walls are usually affected. The line of dampness is usually distinct and approximately horizontal although it is often higher at the corners. The wet area may rise several feet especially if the outside is sealed (eg by a rendering).

Efflorescence from ground salts is almost always present causing decay and damage to plaster and eventually loosening of paint and paper. Hygroscopic salts tend to retain moisture leaving plaster permanently wet.

Mould growth is common on adjacent furnishings and timber in contact with damp areas—eg skirtings, floors, frames—is frequently affected by rot.

Remedial work

Several alternatives are possible, the type of work being determined largely by circumstances such as:

1 Access to walls and working space.
2 Thickness and construction.
3 Disturbance to occupants.
4 Approval for improvement grant—if required.

Insertion of dpc

The original practice of removing one or more courses of bricks in order to build in the dpc is now practically obsolete. The usual method is to cut a narrow slot along a suitable joint bed using a hand saw or power operated reciprocating chain or circular saw or grinding disc. Cutting may be started at a corner or jamb—in the latter by removing a few bricks.

Hand saws are often used at difficult corners and for internal cutting to reduce dust and disturbance. Powered circular and reciprocating saws are faster in continuous work but create dust and noise problems. Power cutting is mainly done externally or in unoccupied buildings.

Dust can be reduced by fitting a cover over the cutting blade and using vacuum extraction, **1.** This is also used with grinding discs.

Depth of cut in each case is determined by the size of blade or disc and is normally limited to 230 mm if done entirely from one side. Cutting is carried out in sections up to 600 to 900 mm long depending on the loading. The damp-proof membrane is inserted immediately, **2,** and the joint space wedged and grouted or pointed with rapid hardening cement mortar.

Good installation practice is shown in **3** to **5.**

Various materials may be used for the dpc; rigid sheets such as copper are easier to insert in thin slots, while low density black polythene is cheap and easily cut and adaptable. Bituminous felt in one or two layers is also used. Adjoining sections of the damp-proof membrane must overlap to provide a continous seal.

Rates of working vary with the thickness and construction of the wall and with the sawing technique but averages about 1·200 m/h for 225 mm and 1·800 m/h for 112·5 mm walls. Damp-proofing a house normally takes three to four days.

1 *Power saw with vacuum dust extractor.*

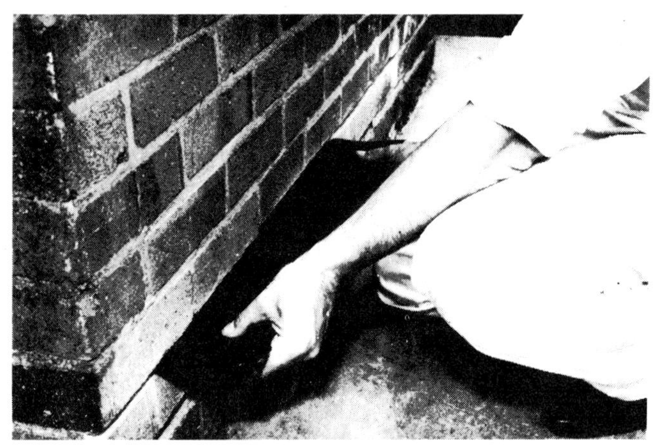

2 *Insertion of damp proof membrane.*

A properly installed damp-proof membrane should have an extensive life and work is usually guaranteed 20 to 25 years.

An important limitation is that a dpc can only be inserted in a coursed wall, ie not in random stone.

Difficulties

In addition to dust and disturbance, difficulties may arise from settlement and wall fracturing.

Settlement is limited by cutting short lengths and by careful packing. It is not normally significant unless the wall is out of plumb or of large expanse.

Fractures and dislodgement of bricks and so on may arise if the wall is badly decayed or mortar is missing from the vertical joints.

Sawing cannot be used without consent in party walls and is difficult in chimney breasts, piers etc. In these situtations chemical damp-proofing may be employed.

The damp-proof membrane must be inserted at least 150 mm above ground level. It must be below the level of any suspended timber floor, or continuous with the dp membrane in a solid concrete floor. This is not specific to mechanical insertion, but applies to any system. In many older buildings this is above the floor and so the dpc should be extended down internally to below floor level, **5.**

Additional works

Additional work may be necessary in all cases of dpc insertion:

1 To replace any rot affected floor timber, **7,** at provide protection to the floor against damp including site concreting and improved subfloor ventilation.

2 To hack off and renew any wall plaster which is severely damp or perished, up to 460 mm above the damp stain. The exposed brick or stonework should be sealed prior to replastering to avoid hygroscopic salts drying out into the plaster. Alternatively the new rendering may contain a water repellent (see BRS Digest 27).

Depending on its condition, the normal rate of drying out for a damp wall is about one day per mm of thickness.

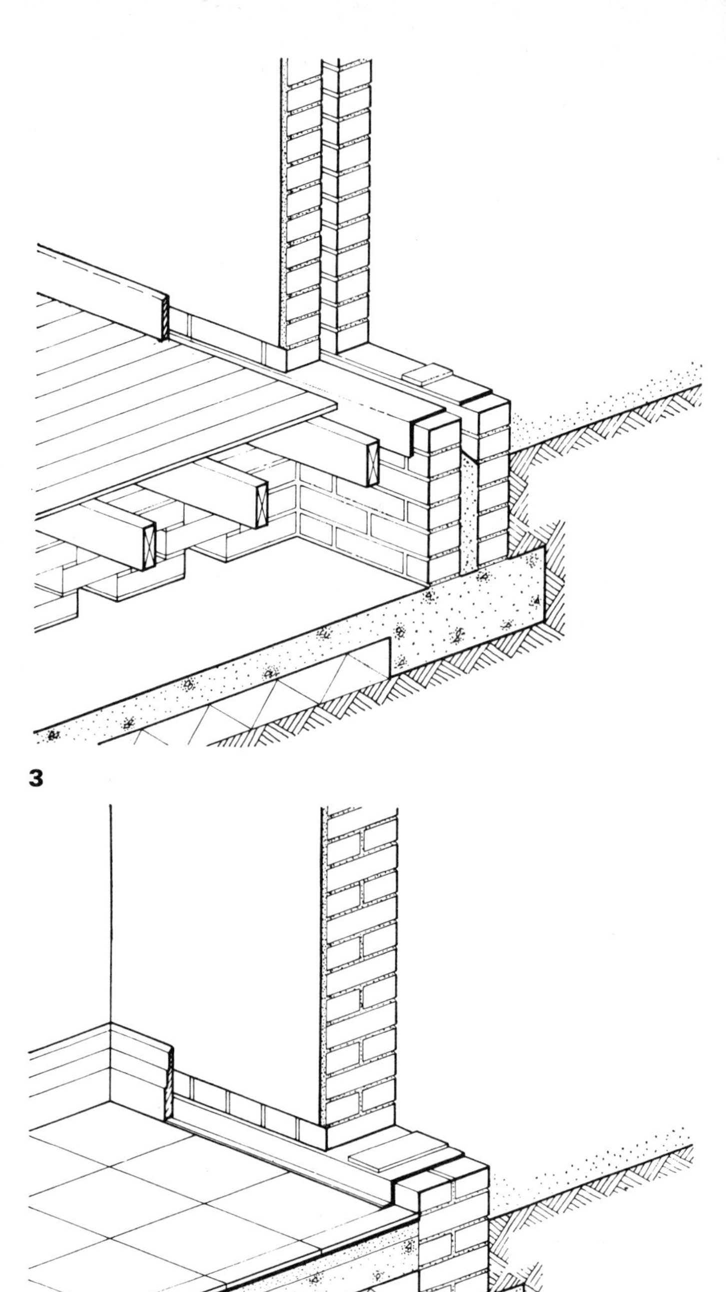

3

4

Chemical treatments

Chemical damp-proofing operates by impregnating the pores of a wall with a solution of water repellent and/or sealing materials. The most common are solutions of silicone resins in white spirit or aqueous siliconate solutions.

A line of holes—about 150 mm apart—is drilled along the base of the wall and the chemicals are injected under controlled pressure or allowed to soak into the material through specially designed nozzles, **6, 8, 9.** Exterior walls can usually be treated from the outside.

Vertical damp-proof membranes can be inserted at wall junctions, Party walls can be treated without damage to the remote side of the wall. Damp-proofing a typical house takes about two days.

Pressure saturation usually carries a guarantee of up to 30 years and other saturation processes 20 years. Failure may result from inadequate saturation, particularly in perished porous mortar, and this is aggravated by voids and excess hygroscopic salts.

To improve drying out, holes must be left open for at least six months. They are not usually left open permanently, though this may happen if the system is subsequently covered over with skirting boards. In the case of rendered/painted walls, Protim fit wall ventilators to provide drainage and ventilation.

Injection is carried out 150 mm above ground level and while the chemicals penetrate down a short distance, additional protection may be required to timber floors below this level (see **5**). Some chemical damp-proofing compositions include a fungicidal agent, though the usual practice is to use a separate fungicide when timber is at risk.

Damp perished plaster will also require renewal and treatment against efflorescence as noted in 'additional works'. Usually costs of chemical damp-proofing are slightly lower than mechanical methods.

Vertical damp proofing techniques

These techniques involve removing wall plaster and providing a vertical barrier to prevent dampness from the wall penetrating inwards. They do not cure rising damp but conceal it within the wall. In all treatments the damp-proof barrier must extend at least 500 mm above the highests visible line of dampness. Because of the sealing effect, dampness has a tendency to rise higher. Types of treatment are given below:

1 Bitumen emulsion or rubber-tar composition applied as a coating film direct to the brickwork then plastered over. This is liable to be damaged by crystallisation of salts accumulating behind the impervious film. Dry shrinkage cracks tend to form in the plaster.

2 Cement rendering using a waterproofing additive. As with bitumen emulsion, salts may loosen and detach the rendering. Salts conveyed by the dampness may react with and damage the cement. Dense renderings often produce surface dampness due to condensation unless a porous finish is used.

Generally these two methods are regarded as repairs rather than improvements and would not normally qualify for improvement grants.

3 Using the same principle impregnated felt in reinforced corrugated sheets may be nailed direct to the wall, **10, 11a, b.** This method is probably most suitable where the insertion of a hori-

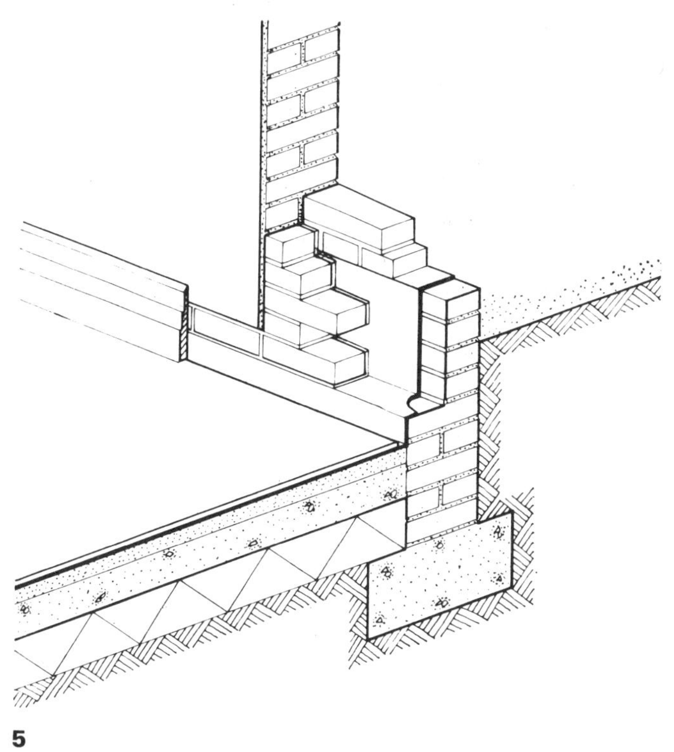

5

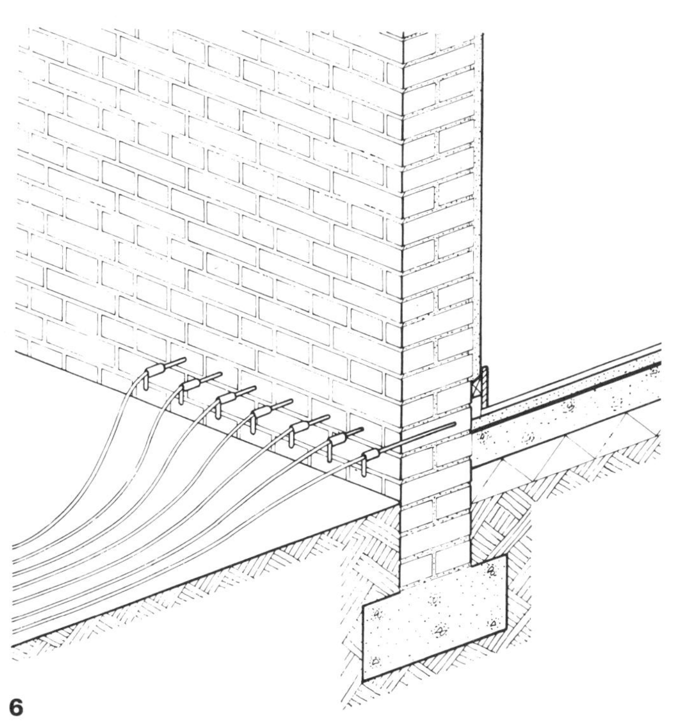

6

7

3 *Dpc must be at least* 150 *mm above ground level. Damp proof membrane should be turned down behind an under skirting or taken down to join floor. Note wedges to prevent wall settlement.*
4 *In cavity walls the damp proof membrane is inserted separately in each leaf. With timber floors, the membrane should be turned down below floor level. Additional treatment may be required to protect floor timbers.*
5 *Stepped dpc required where ground is above floor level. Insertion in existing wall necessitates access to interior of wall and extensive rebuilding. Alternatively, soil can be cut back.*
6 *Chemical injection of silicone—latex compounds under pressure. Wall impregnated through nozzles in drilled holes.*
7 *Rot-affected timber floor.*
8, 9 *Injecting chemical dpc.*

zontal damp-proof membrane is impracticable. Corrugations allow dispersal of vapour pressure behind the barrier while affording a key for plaster.

The damp-proof membrane must extend completely round the wall and behind skirtings. To avoid changes of plaster thickness it is preferable to extend sheeting up to ceiling height. In each case, no protection is given to the floor and this must be considered separately taking into account dampness and rot.

Exterior faces must also be examined. To avoid risk of dampness rising higher as a result of internal sealing evaporation must take place from the external face. Where the external surface is rendered, or otherwise made impervious, the lack of evaporation will cause the rising damp to rise to a higher level.

Porous tube system

The traditional method of drying out a damp wall and, thereby, providing a limit to rising damp is to drill holes and insert porous tubes at intervals along the base 150 mm above ground level. The tubes serve to extract moisture by capillary attraction and are inclined downwards to prevent rain entry and allow drainage while at the same time increasing evaporation.

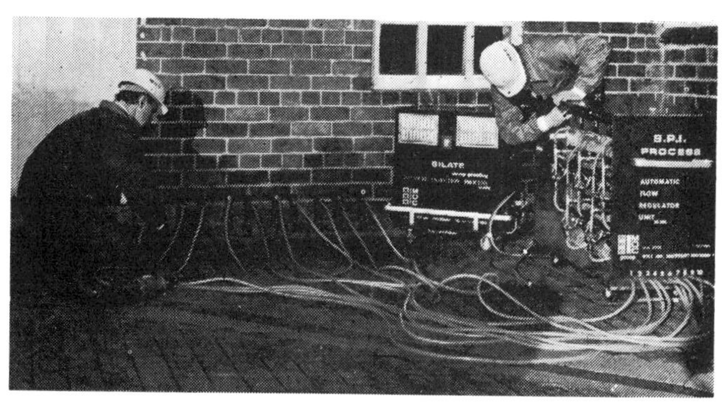

8

9

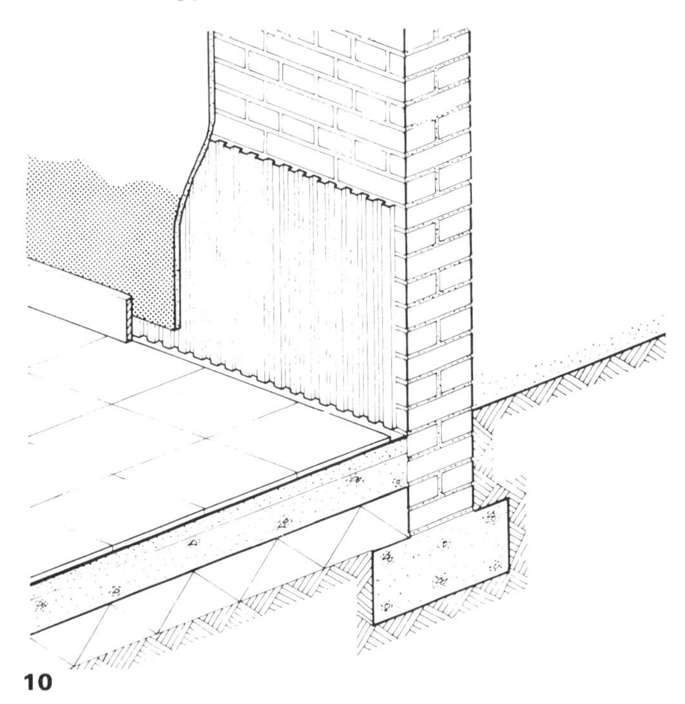

10

11a

11b

10 *Use of bitumen impregnated felt as vertical dpc. Plaster must be hacked off to at least* 500 *mm over highest point of visible dampness. Membrane is then fixed and plastered over.*
11a *Fixing corrugated bitumen impregnated felt.*
11b *Plastering over corrugated bitumen impregnated felt.*

9.3 Damp-proof course by-passed through porous material

Internal bridging

Sometimes the damp-proof course in a wall is bridged by the plaster surface. This usually arises when the wall dpc has not been properly stepped down to join the damp-proof membrane in a solid floor.

It is a constructional fault which becomes evident at an early stage. Plaster at the base of a wall becomes damp and stained near the skirting.

Remedy
The damp-proof course must be located. All wall plaster below this level should be removed with the skirting. A vertical damp-proof membrane is constructed joining those in wall and floor, **12.** This must be firmly keyed to the wall—preferably using liquid bitumen. The wall is subsequently replastered.

External bridging

If an adjoining solid floor or path is constructed above the dpc level in a wall, the wall may become damp as a result of external bridging of the dpc or because of splashing

There is usually a distinct line of dampness along the base of the wall affected. Dampness due to external penetration is more evident following rain.

Remedy
Externally the path should be excavated adjacent to the wall and the wall surface should be bitumen sealed from dpc level to at least 150 mm above ground level. Bitumen may be brushed on to the cleaned wall but must be keyed into the top joint. A cement plinth—also keyed into the wall—is usually necessary for protection, **13.**
Internally the bitumen must join with the floor damp-proof membrane.

Soil

An area of permanent dampness coinciding with areas of walls covered by soil is usually evident and becomes very damp and stained following rain.

Remedy
The ground adjacent to the wall must be excavated to at least 150 mm below the dpc level. An open area (minimum 300 mm wide) with soil sloped back must be provided to ventilate the wall, **14.** The expense of a retaining wall is often justified to keep the area permanently free of soil. In every case the open area must be drained; land drainage may sometimes be necessary.

Cavity failures

Dampness can occur because of mortar and porous material collecting in the bottom of a cavity and filling the cavity space above the damp-proof course level. Similar but more serious effects result from water accumulating in the cavity. Very damp irregular patches form along the base of the wall in the areas affected 15. These are obviously wet following rain and usually begin to show soon after construction.

Remedy
Bricks must be cut out to allow access to the cavity. All mortar filling must be cleared out of the cavity to leave a clear space below the dpc level. Vertical joints in the external leaf 150 mm below the dpc must be left open for drainage. It may be necessary to lower the water table in the area of the site by providing land drainage.

Rising damp in floors

Solid floors of many pre-war properties have no special provision for damp-proofing but rely on the density of the concrete, flags or tiles to prevent penetration of moisture. Inferior bitumen or tar in damp-proof membranes and flooring composition is not an uncommon cause of dampness in properties

built in the immediate post-war period. Damp-proof membranes may also be inadequately sealed during construction or subsequently damaged by settlement disturbance.

Dampness in floors may be temporarily concealed by floor coverings such as linoleum. Eventually a film of moisture tends to collect underneath causing bubbling of the surface and mould on organic materials in floor coverings.

Remedies

Remedial work to prevent dampness penetrating through a solid floor usually involves breaking up and removing the existing floor to enable a new floor to be constructed to the correct level.

In reconcreting (on a suitable hardcore base), damp-proofing may be provided by:

1 Incorporating a water-proofing additive or a damp-proof membrane in the thickness.

2 Providing a damp-proof membrane (eg bitumen) immediately below the floor finish.

3 Forming a damp-proof finish such as mastic asphalt.

In the last two methods the damp-proof layer must adhere to the concrete sub-floor.

Continuity must be ensured between floor and wall dpcs to avoid bridging.

Hydrostatic pressure

In basements, hydrostatic pressure due to water collecting in surrounding ground may lift, fracture or detach the damp-proof layer.

Remedy

Basement damp-proofing requires the whole of the underground area to be enclosed within a continuous damp-proof membrane (tanking). Where leakage occurs this may be prevented by:

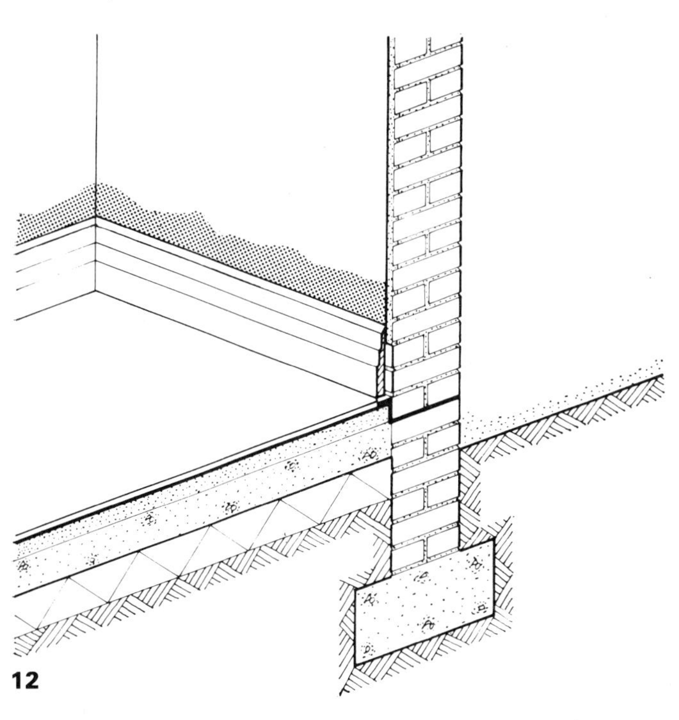

12

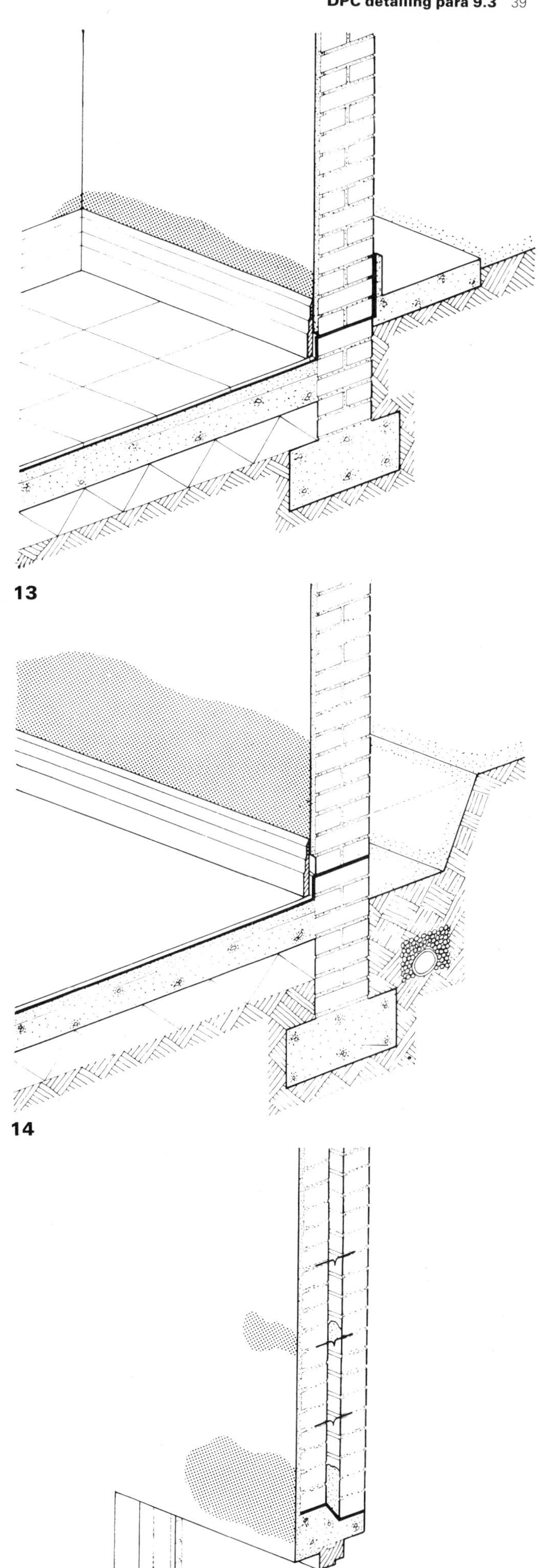

13

14

15

12 *Shaded area shows damp from plaster bridging original badly constructed dpc where wall and floor dpcs were not properly bonded together. Dpc should be reconstructed as shown.*

13 *Shaded area shows damp from pavement constructed over original dpc. Vertical dpc should be added as shown to 150 mm above ground level and protected by cement plinth.*

14 *Shaded area shows damp from soil bridging dpc. Can be corrected by moving soil back at least 300 mm. Land drainage may be necessary.*

15 *Cavity can be bridged by mortar collecting over ties or on dpc over lintels and so on. Bricks must be cut out to remove mortar.*

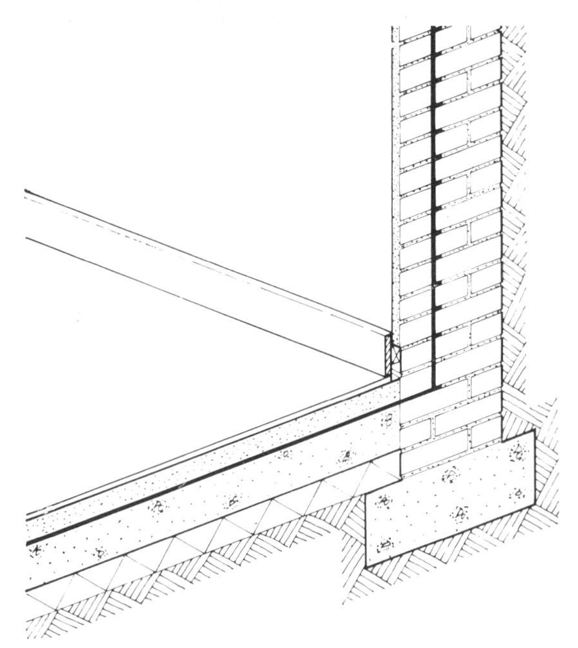

16 *Tanking to prevent penetration of dampness in basement. Vertical damp proof membrane must be constructed supported by brick leaf. Damp proof membrane must be constructed under floor screed. Diagram shows new construction. When adding inner leaf against existing wall, brickwork would rest on floor slab.*

1 Internal rendering with waterproofed cement (the success of this method relies on the adherence of the rendering).
2 Drilling holes and injecting grout behind the basement walls and floor. (It may be difficult to locate the exact point of entry of water.)
3 Providing a complete damp-proof layer supported by an inner wall or floor screed of sufficient strength, **16.**

9.4 Penetrating damp walls

Solid walls: inadequate thickness for exposure

Irregular patches of dampness form during and following heavy rain if walls are not thick enough. This usually affects walls of upper floors exposed to prevailing winds, particularly gable walls. Dampness tends to dry out after a few days of fine weather but may persist longer if plaster is perished.

Remedies
Several remedies may be used depending on the character of the building:
1 Cement rendering on outer surface. The surface is cleaned and joints raked to provide a key. A porous mix (eg 1:1:6 or 1:2:9 of cement: lime: sand, strong mix for greater exposure) is used to increase the absorption capacity of the wall. This is applied in two or three coats, preferable with the final surface rough-cast to reduce cracking and improve water shedding properties. Tyrolean, textured and pebbledash finishes are equally suitable and various proprietary materials may be used. Rendering should not extend over the dpc.
2 Tile hanging and the like. Tiles, usually with 38 mm lap, are nailed to battens, preferably fixed to vertical runners secured on wall. The timber should be creosoted. This method gives good water shedding and ventilating effects. Quoin and edge finishes may present difficulties. Slates, asbestos cement and other materials are also used.

Similar wall cladding systems are used for renovating large buildings with facing panels of various materials either secured to a supporting framework or direct to the wall. Provision must be made in jointing and fixing for lateral movement and drainage.
3 Surface treatment. Provided the surface is otherwise sound, a silicone solution or the like may be brushed onto the wall to reduce capillary attraction; the treatment must be repeated after a few years. This process also helps to preserve the surface of stonework and other materials liable to damage by repeated wetting. It may, however, retard the drying out of dampness from other sources.

Perished joints
Effects are similar to those described in solid walls but more conspicuous patches of dampness occur in wet weather. Joints are visibly open and mortar perished.

Remedy
Perished joints are raked out to a depth of about 25 mm and repointed with suitable mortar. Small areas are completed at a time. Flush pointing is usually adopted. As an alternative surface rendering or tile hanging may be used.

Cracked, displayed and perished renderings
Dense renderings often have shrinkage and movement cracks which may not be obvious. These can cause extensive dampness by capillary attraction of water running down the face of the wall. Impermeability of the rendering tends to prevent the wall drying out. Internally the wall is usually persistently damp with dampness increasing after rain. Large areas of wall are usually affected—often the entire wall—depending on extent of cracking and decay. Defects of this nature are often longstanding and the wall plaster may be extensively perished.

Remedies
Repairs to the cracks may be practicable if the rendering is otherwise sound. Surface sealing and water shedding properties can be improved by painting with cement paint and similar materials. This may need to be repeated from time to time.

If extensively perished, rendering should be hacked off and the exposed wall surface cleaned to remove loose mortar, efflorescence and so on. A new rendering of more porous material, tile hanging or other covering may then be applied.

Ivy, mosses and other vegetation
Vegetation growing over the surface of a wall reduces air flow and evaporation and retains a film of moisture against the surface. Roots usually penetrate and decompose the more porous areas like joints. Mosses and lichens mainly affect northerly walls, particularly in sheltered situations. Gutters and rainwater pipes may become blocked, causing overflow down the wall and extensive wetting.

Remedies
Two types of remedy are available:
1 External treatment. Ivy and similar plants can be destroyed by chemical treatment or by severing the main stems at ground level. Branches are subsequently detached from a wall which invariably needs to be repointed and repaired. Mosses and the like are controlled by spraying with toxic washes. More effective results are obtained by wire-brushing the area first. A wash must be applied in dry weather and further treatment is usually needed at two or three year intervals.
2 Internal treatment. When it is not possible or desirable to alter the external appearance by destroying vegetation. Internal dampness may be controlled by means of a vertical damp proof barrier such as bitumen impregnated sheets fixed to the wall and plastered over.

Penetration around jambs of door and window openings
Damp areas sometimes extend down the sides of window and door openings—mainly on upper windows. Damp becomes conspicuous after heavy driving rain and is caused by the omission of a vertical dpc during construction.

Remedy
The problem may be difficult to remedy. A chase may be cut through the reveal and a vertical dpc inserted. Otherwise jambs may have to be taken down and rebuilt with a suitable dpc in place (providing temporary support to the lintel). If minor, internal treatment may be adequate to conceal the dampness.

Water entry at window and door heads

Water running down the outside of a cavity may drain across door and window lintels to the inside. This is caused by lack of a dpc or inadequate stepping to convey water to the outside. In other cases, there may be no drainage or weepholes may be blocked.

During prolonged rain large quantities of water may drip from the top of a window opening through the lintel or through the adjoining wall plaster. This is usually evident in newly occupied buildings.

Remedy

The brickwork above the lintel must be opened up to inspect the cavity and dpc. Insertion of a dpc is difficult. Depending on width, the work may be carried out in sections to reduce disturbance and settlement but the dpc must be continuous.

Weepholes should be left above the lintel, particularly if strong mortar is used. If the inflow of water is excessive, the wall above should be examined for pointing and structural defects.

Joints between prefabricated panels in curtain walling

Precast and prefabricated panels are normally of impermeable materials. During driving rain water streams down the outer face of the building and tends to be channelled into jointing grooves and it may enter through unsealed or defective joints and cracks caused by differential movement (thermal, settlement, moisture, wind pressure). Where joints allow rain entry, weepholes and drainage channels may be blocked.

Water penetrating into a cavity or insulation space runs down until deflected by a frame, beam or the like. Large quantities of water may drip down the inside of walls. It may also travel inwards across impermeable structural members (eg concrete floors) or along service chases to emerge some distance away.

Table XI Agrément certificate: Damp proof courses: walls—chemical injection*

Certificate	Company	Type
No 79/642 Peter Cox DPC Process	Peter Cox Ltd Wandle Way Micham Surrey CR4 4NB	Silicone transfusion
No 79/673 Permadry	Permadry Ltd 48A Walsall Road Four Oaks Sutton Coldfield Warwickshire B74 4QT	Silicone transfusion
No 80/725 Catalpa	Catalpa Chemicals Ltd 75 Market Street Pocklington York YO4 2AE	Stearate injection
No 81/824 Rentokil	Rentokil Ltd Felcourt East Grinstead Sussex RH19 2JY	Silicone injection
No 81/825 Nubex	Tenneco Organics Ltd Rockingham Works Avonmouth Avon	Silicone injection
No 81/827 Freezteq	Norman Rudd (DP Division) Ltd 110 London Road Aston Clinton Aylesbury HP22 5HS	Silicone diffusion
No 81/833 Wyamkit SPI **No 81/834 Wyamkit PI**	Wykamol Ltd Tingewick Road Buckingham, Buckinghamshire MK18 1AN	Stearate injection Silicone injection
No 81/840 Ness	Thomas Ness Ltd Eastwood Hall Eastwood Nottinghamshire NG16 3EB	Silicone injection
No 81/862 Triject 2	Triton Chemical Mfg Co Ltd Lyndean Industrial Estate 145 Felixstowe Road Abbey Wood London SE2	Silicone injection
No 81/872 Mystosil DP	Catomance Ltd 88-96 Bridge Road East Welwyn Garden City Hertfordshire AL7 1JW	Silicone injection
No 81/873 Manalox	Manachem Ltd Ashton New Road Manchester M11 4AT	Stearate injection
No 81/874 Permoglaze	Blundell Permoglaze Ltd Charnley Fold Lane Bamber Bridge Preston PR5 6AA	Stearate injection
No 81/875 Stanhope DPC No 1	Stanhope Chemical Products Ltd 37 Broadwater Road Welwyn Garden City Hertfordshire AL17 1JW	Stearate injection
No 81/885 Kingston 8	Pentatherm Ltd 12-13 St Clements Oxford OX4 1YR	Silicone injection
No 81/896 Rentokil	Rentokil Ltd Felcourt East Grinstead Sussex RH19 2JY	Silicone injection
No 82/967 Remtox Silicone **No 82/968 Remtox Stearate** **No82/969 Remtox Silicone AQ**	Remtox Chemicals Ltd Howison Court Gillingham Dorset	Silicone injection Stearate injection Siliconate injection

* See table X for flexible dpcs with Agrément certificates.

Agrément certificate: Damp proof courses: foundations

Certificate	Company	Type
No 76/397 Tretolastex **No 79/685 Tretoflex**	Tretol Building Products Ltd Tretol House Edgware Road London NW9 0HT	Bitumen- rubber membrane Pitch polyurethane
No 79/675 Sika-1	Sika Ltd Watchmead Welwyn Garden City Hertfordshire AL7 1BQ	Silicate additive
No 80/723 Visqueen 1200 Super	ICI Plastics Division Visqueen Products Six Hills Way Stevenage Hertfordshire SG1 2DB	Polyethylene
No 80/750 Bituthene Standard and 1000	Servicised Division of W.R. Grace Ltd 2 Caxton Street Lodnon SW1H 0QJ	Polyethylene
No 80/754 HT 350	D Anderson & Sons Ltd Barton Dock Road Stretford Manchester M32 0YL	Polyester fibre/ bitumen tanking
No 80/757 Liquapruf	Colas Products Ltd Galvin Road Trading Estate Slough SL1 4DL	Bituminous emulsion
No 80/769 HydrEpoxy 300	Unibond Ltd Tuscam Way Yorktown Industrial Estate Camberley Surrey GU15 3DD	Resin emulsion
No 81/836 Hansit Waterproofer	Hansit Exporters Ltd 8 Walker Street Edinburgh EH3 7LA	Cementitious compound

Remedies
Water entry points must be located; this may be difficult. Tracing indicators such as dyes are sometimes used by this must be done with care to avoid staining and subsequent damage. The extent and nature of the work depends on the materials and condition of the wall. It may be necessary to dismantle sections to renew panels or refit 'dry' joints. In other cases, defective joints are cut or cleaned out and remade with suitable sealant or mastic applied under pressure.

Defective window and door frames
In driving rain, areas of reveal may become wet due to failure of pointing around a frame.

Remedy
The joint between the outside of the frame and the wall jambs must be repointed with mastic applied under pressure.

Defective window sills
An area of porous wall below a window sill may become saturated by water draining from the window. This usually arises from ineffective drip grooves and water bars and lack of suitable damp-proof treatment.

Remedy
The sill should be examined and repaired, reformed or renewed as necessary. Brickwork below the sill may have to be cut out to insert a dpc. Dampness below windows may also result from water splashing from the ground and other ledges.

9.5 Roof problems

Penetration through defective parapets
A common source of dampness in upper rooms is caused by perished pointing, broken copings, defective dpc flashings and so on in parapets. Rain entry at this point—unless arrested by a dpc—quickly travels down through the vertical joints and voids in a wall. Water may also spread into the porous interior of adjacent flat roofs, resulting in insulation becoming saturated with increased heat loss and risk of condensation in the roof. Water may travel across structural concrete to emerge some distance away.

Internal dampness usually spreads down a wall surface and across the adjacent ceiling. A ceiling is usually more affected if leakage comes through flashings or occurs at the junction with the roof. Dampness is most evident following heavy rain but usually persists for several days when the wall becomes saturated and may be almost permanently damp in winter.

Remedies
A full examination of the parapet and associated roof drainage is necessary. Depending on its condition, remedial works may include raking out and repointing open and perished joints, cutting out and rebuilding decayed and defective brickwork, renewal and resetting of broken or perished copings, renewal and refixing defective flashings.

If extensively perished or dangerous the parapet must be taken down and rebuilt in sound materials. Materials used in this exposed position must be durable (eg engineering bricks in cement mortar 1:3).

Correct positioning of cover flashings and dpc is very important, **17, 18.** Dpcs must be provided under copings to prevent water penetrating down through the joints between coping sections.

If the parapet is low, the impermeable covering to the roof or gutter may be extended up and over the parapet (below the coping) as a continuous barrier. Otherwise the roof covering or gutter lining is extended up at least 150 mm and overlapped with a cover flashing. A second dpc is provided at this level to prevent water by-passing through the wall, **17, 18.**

Rain penetration through chimney stacks
Dampness from defective chimneys is similar to dampness from defective parapets but spreads down the chimney breast and into the adjacent ceiling. Staining and efflorescence may be caused by carbon and acid salts leached from the flues.

More persistent dampness may occur in disused flues which have not been sealed or sealed without means of through ventilation. In both cases, dampness may also arise from condesation and from hygroscopic salts absorbing moisture during cold damp weather.

Remedies
If stack brickwork is sound, repointing and reflaunching may be

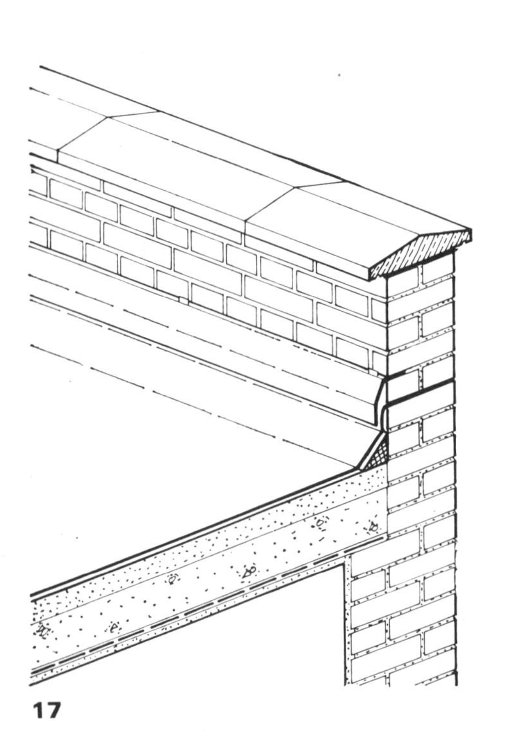

17

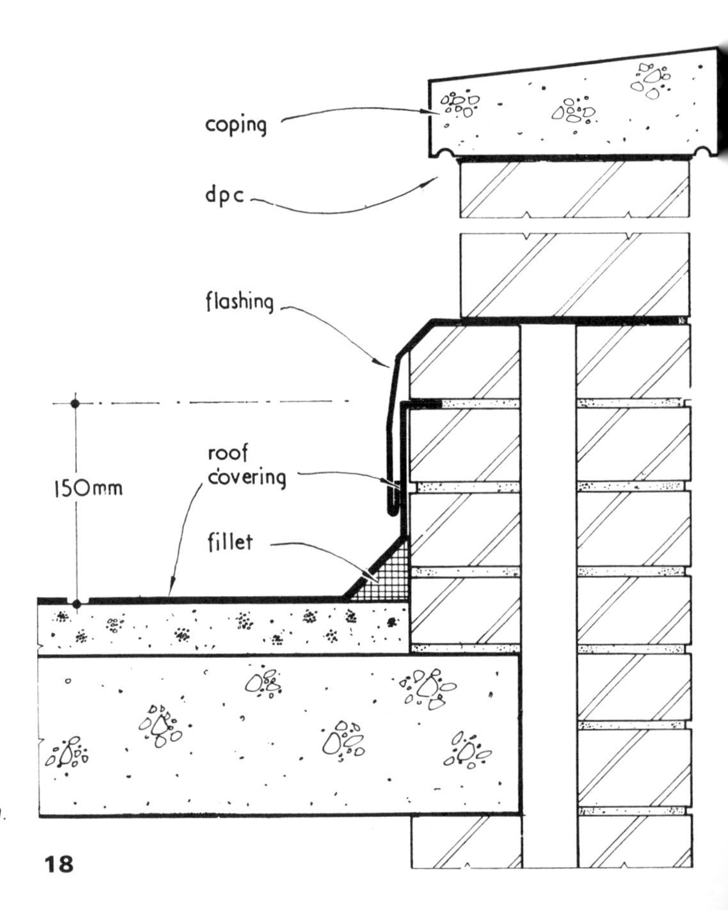

18

17 *Roof and parapet junction (for other examples, see p14).*
18 *Typical dpc and flashing details at roof and parapet junction. Fillet can be made of any suitable sterile material (eg expanded polystyrene or cement).*

satisfactory. Brickwork of chimney stacks is often damaged and distorted by sulphates from the flue gases attacking the building materials. If extensively perished, bulging, leaning or otherwise dangerous, stacks should be taken down and rebuilt. Durable materials must be used in reconstruction; flashings must be renewed where necessary **19.**

Disused stacks may also be taken down— preferably to roof level and roofed over. Alternatively, a ventilating cover should be provided at the top and bottom of each flue. It is important to prevent birds getting into closed flues.

Leaking and overflowing rainwater pipes and gutters

Water discharged down the face of a wall from leaking pipes and gutters tends to wash out joints and cause severe local dampness 20. Splashing may cause dampness at base of a wall and on walls adjacent to ledges and cornices. Overflows may be due to silt, moss balls and so on blocking pipes or gutters or they may be due to indadequate falls or gutter supports. Leaking rainwater pipes may have open joints or cracks—often these are inconspicuous down the side nearest the wall. In all cases, wetting is evident during and after rain. In cavity walls the water may penetrate across bridging faults (see cavity failures).

Remedies

Pipes and gutters should be examined during rainfall. Cracked pipes can be identified by a dull grating note on tapping; they must be replaced. Sections of gutters may need to be renewed, realigned to falls, rejointed and/or firmly secured.

Characteristic properties of materials give clues to types of defect most likely to arise, eg cast iron (rusting, breaking, cracking), asbestos cement (breakage), pvc (detachment, leaking joints).

Leakage through sloping roofs

Broken and displaced slates and tiles allow rainwater to drip into areas of ceiling or wall giving rise to distinct wet patches during rain. Dampness is usually temporary. The most serious leaks often occur near the eaves where water may also penetrate down the inside of the wall face and at valley gutters where water flow is concentrated.

Roof defects may arise from decay, wind leverage and breakage from access and objects falling from adjacent structures. A slated roof may appear to be sound superficially while the slates may be decayed and porous under the lap. Slate pegs and nails may decay and corrode allowing slates to slide out of position if moved by the elements. Similarly, nibs of tiles may be decayed by salt crystallisation. Wind dislodgement of hip and ridge tiles is common and these must be bedded and pointed in cement mortar. Slates and tiles at eaves and verges are also liable to wind damage and, if loosened, may lead to large areas of roof being stripped.

In older property, roof timbers may be warped, rotted, worm-eaten and/or broken producing distortion. If rafters are affected this is shown by sagging between supporting walls. Damp may appear temporarily if there is no sarking beneath the roof; wind blown rain and snow is able to collect below tiles and drip on to the ceiling.

Remedies

Access to a roof is necessary for a detailed assessment. Individual slates and tiles can usually be replaced by sliding them into position. Slates are retained in place by lead tacks. Tiles are levered so that the nibs hook over the battens. Near eaves and verges, tiles must be nailed and properly bedded with cement mortar.

If extensively broken and decayed, the whole area of roof should be stripped and retiled or reslated. Provision is usually made for the re-use of 50 per cent of slates if they are substantially sound.

If the roof is badly distorted it is usually necessary to renew roof timbers. In all cases, the specification should include for the replacement of sarking and associated flashings, soakers and gutters where necessary.

Leakage through flat roofs

A relatively small hole or perforation in the impermeable covering of a flat roof can admit large quantities of rainwater. The most serious dampness usually arises from leaks in channels and drainage gutters.

On a porous base, eg lightweight concrete, penetrating water may be absorbed and spread over a large area. In this case damp will tend to persist for some time following rain. Rainwater reaching dense structural concrete may drain along conduit holes and form drips some distance away. In metal and felt covered roofs, water may drain towards and down the inside of adjacent walls.

Faults giving rise to leakage include:

1 Mechanical damage from access or wind.

2 Cracking due to thermal and moisture changes.

3 Blistering as a result of vapour accumulation under the covering.

4 Gradual corrosion and decay.

Dampness may also penetrate through the associated flashings (see parapets and flashings).

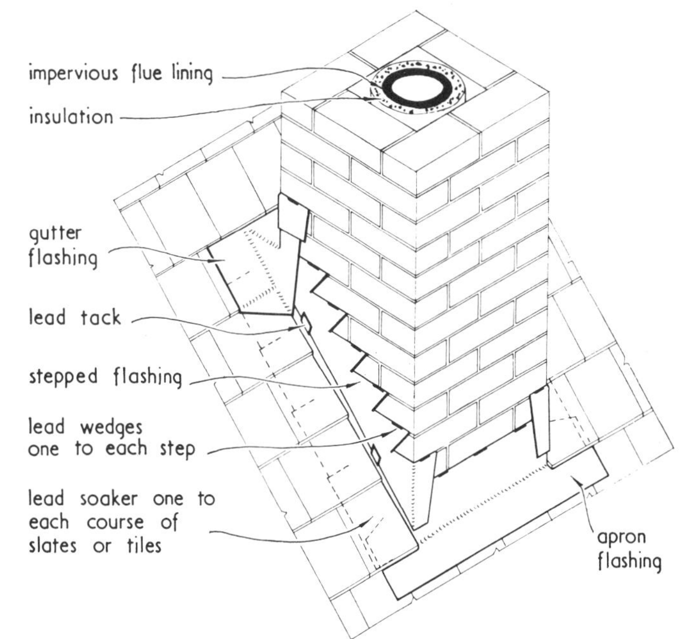

19 *Lead flashing details at junction of chimney and roof.*

20 *Decayed brickwork caused by defective rainwater pipe.*

Remedies
The extent of remedial work depends on the nature and condition of the roofing material. Small perforations in metal roofs are normally repaired by soldering. If a larger area is damaged whole pieces of metal may have to be taken out and replaced.

Concrete roofs may be resurfaced with mastic asphalt. Repairs are often effected with liquid bitumen in emulstion or solution which can be painted directly on to the surface. In each case roof surfaces should be finished with reflecting materials (eg white or aluminium paint or limestone chips). Roofing felt may also be repaired with bitumen but, if extensively defective, it is more satisfactory to replace the felt completely.

In roof repairs it is usually necessary to renew or refix cover flashings and the like and this should be provided for in the specification.

Entry of rain through flashings
Where a wall or other structure extends above the level of a roof, provision must be made for junction to be effectively covered. Dampness may arise from bad arrangement or defects in flashings, soakers, valley gutters and dpcs. If these are unsuitably positioned rain may enter due to splashing, overfilling and capillary action. Dampness may by-pass damp proof barriers to penetrate down walls or extend into adjacent roofs. Flashings may be torn or dislodged by wind or thermal movement.

Metals undergo gradual corrosion which is greatly accelerated by the sulphurous and sulphuric acids discharged in flue gases. Rapid galvanic corrosion can result from the use of dissimilar metals eg lead and zinc in contact.

In older properties, cement fillets may be used to cover the junction instead of flashings. Cement mortar tends to crack, shrink and separate from the roof and wall.

Remedies
Flashings, aprons and other covers may need to be repaired and refixed or renewed. The top edge of the flashing or upstand must be carefully recessed into a mortar joint at least 150 mm above the roof level and secured with wedges and mortar.

Gutters may need to be repaired or relined and regraded to falls.

The roof covering should be dressed over a fillet and extended up walls. Roof coverings of asphalt and bitumen are carried up 150 mm and the top edge is recessed into a mortar joint for sealing. In all cases, the edge of the upturn must be adequately covered by a flashing. A dpc should also be provided at this point to ensure a continuous barrier, **18.**

Materials used for flashings include: lead, copper, aluminium pitch impregnated asbestos sheeting and plastic materials. The last two can be obtained in preformed sections or can be heat formed on site. Aluminium should have a bitumen coating to prevent contact with mortar or cement. Zinc may also be used but has a limited life unless suitably protected (eg by bitumen).

9.6 Plumbing failures

Burst pipes usually begin to leak in a thaw following a severe frost. The surrounding area quickly becomes wet. Smaller perforations—arising from corrosion—may occur in pipes built into the structure, eg in hot water pipes in the fire back and flue brickwork. In this case, leaks may not be immediately obvious.

Leaks may also develop in cold water cisterns, hot water cylinders and in sanitary fitments. Water may seep into the surroundings for a long period and cause considerable damage due to decay. Condensation on the cold surfaces of cisterns and pipes is very common and may drip onto other areas (see condensation on plumbing).

Remedies
Water pipes, tanks, and so on must be periodically examined for leakage. Persistent dampness in one area may be due to a burst and tests should be applied to pipes in this position. Leaks may also occur in a hot water heating system—particularly in the joints and connections. Suitable means of access for inspection should be provided where pipes are hidden below floors or in other enclosures.

In rooms in which water is used and some splashing is unavoidable (eg kitchens, bathrooms, process rooms), the surrounding surfaces should be impervious and inert to water. If large quantities of water are involved provision must be made for floor drainage.

9.7 Constructional moisture

Large quantities of water are introduced to a building during construction and take a considerable time to dry out fully. In the initial drying period, surface moisture evaporates quickly. Removal of water from the interior is a slower process and relies mainly on diffusion of water vapour through drier outer layers.

During this stage, salts from building materials are carried in solution to the surface and left as efflorescence as water evaporates. Efflorescence and cryptoflorescence* may persist if a material containing salts is repeatedly wetted or if water containing salts is introduced from other sources.

The alkaline properties of concrete and mortar are destructive to oil based paints and these should not be applied until an element is thoroughly dried or a sealant is used.

While a structure is still saturated, its heat conductivity is considerably increased. This may cause condensation on the surface or within interstitial spaces.

Remedies
Evaporation may be accelerated by heat and ventilation. A porous surface dries more rapidly than one which is dense and impermeable. During this period, sealing of a surface (eg by painting) prolongs drying out and the surface covering may blister or suffer other damage.

With concrete roofs which are sealed on the upper surface, water tends to become trapped in the insulation layer or screed covering the structure. Sealing is even more pronounced if a vapour barrier is used on the underside.

To enable trapped water to escape and to ensure that water penetrating the roof structure in future can dry out, porous interiors of roofs should be ventilated. This is effected by forming small ducts in the concrete connected to ventilator outlets on the surface or at the sides of the roof **21a, b, c.**

Similar provisions should be applied to walls and so on which have an impermeable outer surface.

9.8 Condensation

Condensation is liable to occur in many situations, both on surfaces and within interstitial spaces of a building structure. It is essentially related to two main factors:

1 Internal water vapour pressure—which depends on external conditions, internal sources of moisture and ventilation rates.

2 The heat and vapour transmitting properties of the materials forming the building structure.

Condensation is liable to form if:

1 Internal temperature and humidity are high.
2 External temperature drops very low.
3 Heat is rapidly lost through the building structure.
4 Vapour is retained within the structure.

Remedies
Remedies for condensation depend on circumstances and may include:

1 Ventilation to remove steam and excessive quantities of water vapour at source.

2 Background heating to maintain a more or less even temperature within the building at all times.

3 Good insulation to reduce heat loss and to ensure that the

* Cryptoflorescence: formation of crystals within interstitial spaces of building materials eg plaster. Materials are gradually broken down by mechanical pressure and by hygroscopic effects.

temperature of the structure is kept above the dewpoint of the vapour passing through.
4 Vapour barriers to minimise build up of vapour pressure within the interior of the structure. This is important if the outer surface is sealed and if the inside surface is insulated.

Bathrooms and kitchens
Temporary surface condensation results from steam and evaporating hot water.

Remedies
Condensation can be reduced by extract ventilation of the humid air and improved insulation of walls and ceilings. These should be sealed against vapour entry.

Unheated rooms (eg bedrooms)
Condensation of warm humid air occurs on colder wall and ceiling surfaces. This often results in mould growth—particularly in the corners of external walls and behind furniture.

Remedies
Some background heating is desirable. Doors should be kept closed to minimise entry of warmed, humidified air from other parts of the building. Mould growth must be treated with a fungicide before redecoration of the wall. Surface insulation is usually effective in preventing condensation on wall surfaces.

Intermittently heated rooms
In rooms such as living areas which are only used in evenings, the internal climate may become hot and humid as a result of household activities, while the building structure remains relatively cold.

Remedies
Some constant background heating is desirable. Walls may be lined with insulation and sealed. Steam must be extracted from communicating kitchens.

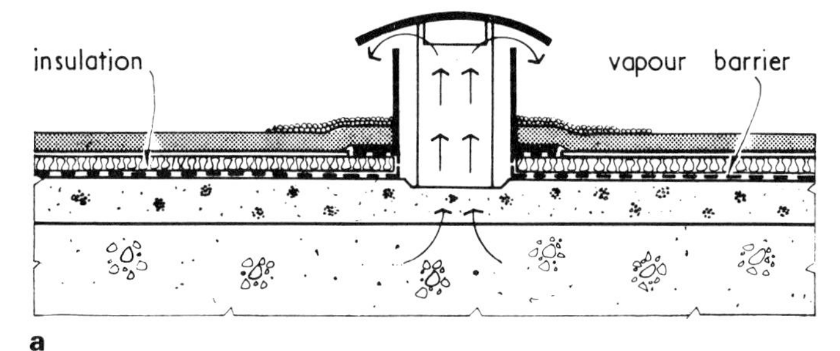

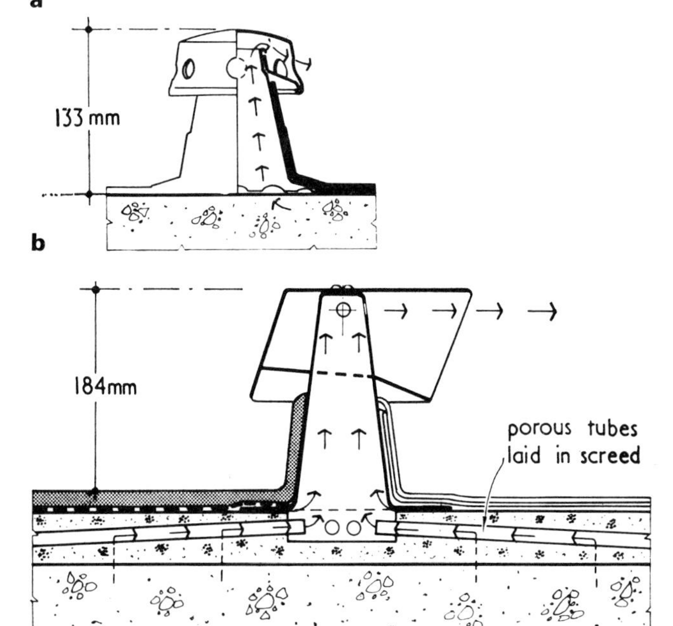

21a, b, c *Forms of roof ventilation for extracting construction moisture.* **21c** *includes porous pipes laid in screed to convey moisture to ventilator.*

Windows
Due to the high conductance of thin glass, inner surfaces of windows cool rapidly (eg at night) to below the dewpoint temperature of room air.

Remedies
Condensation is reduced by siting radiators under windows and promoting a flow of warm air over the glass. Double glazing is very effective provided the space between the panes is sealed. As a palliative, sills may be grooved and drained to remove condensate.

Isolated areas, corners of frame buildings
Condensation may be due to excessive cooling through heat bridges formed where the structural frame extends through the cavity and/or insulation.

Remedy
This is a fault in design and can only be remedied by increasing the insulation at heat bridges.

Enclosed spaces
Condensation may arise from inherent dampness, lack of ventilation and heat in enclosed spaces such as cupboards.

Remedy
Ventilation should be provided and surfaces sealed to reduce dampness.

Chimney breasts
In disused chimneys condensation results from cooling due to dampness (high conductivity and loss of latent heat) and lack of insulation. Hygroscopic salts may also tend to retain moisture. Flues on external walls give most problems.

Remedies
The chimney flue should be ventilated and the top covered to prevent entry of rain. Internally, the chimney breast may be lined with insulation to reduce heat loss. If dampness is great, it may be due to a defect in the chimney, eg lack of a dpc (see para 9.5).

Within chimneys
Condensation may result from low temperature combustion of damp solid fuels in modern stoves or from combustion of oil and gas at very high efficiencies using a minimum of excess air. Steam produced tends to condense on chimney sides causing staining and dampness.

Remedy
Flues should be lined with impervious materials and provision should be made for draining condensate. This liquid is very acidic and destructive to building materials.

Within porous interiors of concrete roofs
Condensation occurs in concrete roofs due to water vapour accumulating below the impermeable roof covering under cooling conditions. This also reduces the insulation properties and tends to produce further condensation.

Remedy
The underside of roofs should be sealed with a vapour barrier. To allow effective drying, the interior of the roof should be ventilated by means of small ducts extending to roof ventilators (see **21a, b, c**).

Below sloping roofs
Condensation occurs below roofs particularly where the roof covering has a high heat conductivity (eg metal and asbestos cement sheeting) and internal humidity is high. If the slope is sufficient, condensate tends to drain down towards the eaves but it may form drips at purlins and other obstructions. Water dripping on to the ceiling or linging causes staining and areas of dampness.

Remedy
In theory, insulation should be placed above the roof to retain heat in the covering. In practice, insulation may be placed below the roof provided the lower surface is sealed with a vapour barrier. The space above the insulation should be ventilated to the outside. In domestic buildings, ventilation of roof spaces is adequate to prevent condensation.

In subfloor spaces
The space below a ground floor normally has high humidity unless it is well ventilated. If hot pipes and so on are in contact with any part of eg damp sleeper walls or site concrete, evaporated water increases humidity and tends to cause condensation on cooler parts of the structure. Similar effects may be caused when hot air ducts in subfloor construction are not effectively damp-proofed. In this case the humid air is conveyed into the rooms of the building.

Remedies
Heating pipes must be well lagged and should not be fitted in direct contact with a damp structure. Subfloor space must be thoroughly ventilated with ventilators on opposite walls. Hot air ducts should be insulated and the outside of the insulation must be surrounded by a damp-proof layer at any place where there is contact with a wall or floor below dpc level.

Condensation on plumbing
Condensation is caused by the rapid cooling of warm air when it comes into contact with cold water cisterns, pipes and so on. Condensate may drip on to other areas and cause staining. If surfaces are covered with an absorbent lagging, the lagging material may become saturated with condensate. In areas in which there is a high humidity the insulating cover should be non-hygroscopic or enclosed within a waterproof cover.

Remedies
Appropriate thicknesses of insulation can be determined for particular conditions (see AJ Services handbook). Insulating materials used for this purpose should be inert to water and non-hygroscopic. The underside of a cistern is often left open to prevent freezing. In this case, a drip tray should be provided or the ceiling must be vapour sealed to minimise the quantity of water vapour coming into contact with the cold surface. This is particularly important over bathrooms and the like in which large quantities of steam are produced.

9.9 Costs of remedial work

Other than in very general terms it is difficult to quote costs of repairs because of lack of consistency in the work. Travelling and other overheads are often disproportionately high and there is a variation in establishment charges.

The following unit prices are typical of a medium-sized job in which the work is reasonably straightforward. Allowances have been made in material costs for small quantities, cutting and waste and for the extra labour involved in jointing to the existing work and making good. A total of 30 per cent on the basic costs has been allowed for profit, establishment and site overheads. The figures are based on material and labour costs current in the London area at September 1982 and should be used as guide prices only. No allowance has been made for VAT.

Outline of work	**Unit prices** Unit	Price £
Damp-proofing walls		
Cut out mortar joint along base of wall, supply and insert polythene or bitumen felt damp proof membrane and wedge and point (excluding plaster work)		
half brick wall	metre	**14.00**
one brick wall	metre	**26.00**
one and a half brick wall	metre	**40.00**

Outline of work continued	**Unit prices** Unit	Price £
Damp-proofing walls		
Drill existing wall and provide chemical dpc by injection under pressure (excluding removal and reinstatement of plaster)		
half brick wall	metre	**8.00**
one brick wall	metre	**13.00**
one and a half brick wall	metre	**19.00**
Drill existing wall and provide chemical dpc by infusion by gravity (excluding removal and reinstatement of plaster)		
half brick wall	metre	**6.00**
one brick wall	metre	**9.00**
one and half brick wall	metre	**15.00**
Drill 75 mm holes in existing external wall and insert porous ventilator tubes and make good		
one brick wall	metre	**25.00**
Hack off and remove damp wall-plaster, rake out joints to form a key, render with aerated water-proofed cement and set; including joint to existing plaster	metre^2	**11.00**
Hack off and remove damp wall-plaster, fix vertical bituminious felt damp-proof membrane and replaster, including joint to existing plaster	metre^2	**14.00**
Damp-proofing floors		
Break up concrete floor, excavate over area, spread and consolidate 150 mm hard-core and lay new water-proofed concrete floor, 115 mm thick, screeded to receive flooring and trowelled smooth to level	metre^2	**22.00**
Take up quarry tiles, apply water-proof primer to surface and make up levels with cement and sand (max 25 mm) and apply one coat latex cement to receive floor tiles	metre^2	**12.00**
Supply and lay thermoplastic tiles on prepared surface of floor	metre^2	**5.00**
Supply and lay thermoplastic tiles on prepared surface of floor	metre^2	**12.00**
Take up rot-affected timber of ground floor and destroy: clean out sub floor cavity and supply and fix new wall plates, 50 mm×100 mm joists and 25 mm t and g boarding appropriately treated with preservative	metre^2	**20.00**
Repairs to defective brickwork		
Hack off defective rendering from external walls, prepare for sand and cement render 2 coats (plain face) (excluding scaffolding)	metre^2	**8.00**
Rake out mortar joints to brickwork and repoint (excluding scaffolding)	metre^2	**6.50**
Rake out mortar joints to brickwork of chimney stack and repoint (excluding scaffolding)	metre^2	**9.00**
Cut out brickwork in small areas, clean out cavity, rebuild and make good	metre^2	**50.00**
Cut out crack in brickwork and make good with new common bricks and point to match existing		
half brick wall	metre	**28.00**
one brick wall	metre	**53.00**
one and a half brick wall	metre	**76.00**
Rake out defective pointing around door and window frames and repoint with mastic	metre	**2.00**
Take out existing stone or concrete sill and supply and build in a 225 mm×75 mm cast concrete sill including making good	metre	**35.00**
Supply and fix creosoted battens plugged to wall at 100 mm gauge with 265 mm×165 mm plain concrete tiles nailed every course	metre^2	**26.00**

Outline of work continued	Unit prices Unit	Price £
Roof and chimney work		
Take off existing coping, cut out and renew perished bricks and supply bed and point a 300 mm×50 mm cast concrete coping; including bitumen dpc built into wall	metre	**19.00**
Take down chimney stack 680 mm×680 mm×900 mm high above roof to 300 mm below roof level (excluding scaffold)	each	**62.00**
Extra for each additional 300 mm of height of chimney	each	**17.00**
Rebuild chimney stack 680 mm×680 mm×900 mm high above roof line with new common bricks in cement mortar, including provision of dpc at roof level and for parging and coring (flashing extra) (excluding scaffold)	each	**120.00**
Extra for each additional 300 mm of height of chimney	each	**29.00**
Extend and construct roof over area of chimney, including new rafters, battens, felt and slates or plain tiles suitably lapped and bonded to the existing roof	metre^2	**85.00**
Take off the exisiting chimney pot, supply set and flaunch a new 0·45 m pot	each	**26.00**
Scaffolding up to two storeys (10 m height) to provide working access for demolition and rebuilding of chimney stack		**150.00**
Cut out defective brickwork and replace with new common bricks and point to match existing in small areas		
half brick wall	metre^2	**49.00**
one brick wall	metre^2	**92.00**
Strip roof slates, sort and reslate roof using 50 per cent of existing slates including the stripping and renewal of roofing felt	metre^2	**20.00**
Strip slate battens, remove nails and renew battens 216 mm centres over an area exceeding 8 m^2	metre^2	**4.00**
Take off and renew up to 10 single 508 mm×254 mm slates including clips	each	**5.00**
Strip plain tiles from roof and retile using 50 per cent of the existing tiles; including the stripping and renewal of roofing felt 265 mm×165 mm tiles	metre^2	**19.00**
Strip tile battens, remove nails and renew battens to 100 mm centres over an area exceeding 8 m^2	metre^2	**4.00**

Outline of work continued	Unit prices Unit	Price £
Roof and chimney work		
Take off and renew up to 10 single plain tiles	each	**3.50**
Take up and redress, wedge and point lead flashings over slates and tiles	metre	**3.50**
Take up and relay the existing lead sheet to boarded flat roof including dressing over drip and rolls and forming new bossed ends	metre^2	**58.00**
Seal cracks and holes in asphalt flat roofs and apply two coats of bitumen	metre^2	**4.00**
Take down existing cast iron gutters and provide, joint and secure new gutters laid to falls	metre	**10.00**
Rainwater gear		
Take down existing rainwater pipes and fix new pipes complete (63·5 mm) diameter	metre	**14.00**
Clean out gutters and pipes and remove rubbish	metre	**0.55**

References, section 9

BRITISH STANDARDS INSTITUTION

1 BS CP: 102:1973 Protection of buildings against water from ground

2 BS 743:1970 Materials for damp proof courses

3 BS 3826:1969 Silicone-based water repellants for masonry

BUILDING RESEARCH ESTABLISHMENT

4 Digest No 27 (Second series) Rising damp in walls. 1969, HMSO

5 Digest No 110 (Second series) Condensation. 1972, HMSO

6 Digest No 152 (Second series) Repair and renovation of flood damaged buildings. 1973, HMSO

7 Digest No 163 (Second series) Drying out buildings. 1974, HMSO

8 Digest No 180 (Second series) Condensation in roofs. 1975, HMSO

9 Digest No 217 (Second series) Wallcladding defects and their diagnosis. 1978, HMSO

10 BRE 245 Rising damp in walls: diagnosis and treatment, HMSO

11 BRE News 55, Winter 1981, Cornish J. P., 'Condensation and mould growth', HMSO

12 BRE News 57, Summer 1982, Newman A. J., Whiteside D and Kloss P. B., 'Water penetration tests on twelve cavity fills', HMSO

10 Reference

1 Lyall Addleson *Building failures* Architectural Press 1982. Revised and expanded version of *Architects' Journal* series.
Contains remedial advice and valuable references on dpc detailing.

2 The Agrément Board *Information sheet No 10* May 1977.
Sets out calculation of 'total exposure factor E' which was adopted in BS 5618: 1978.

3 Brick Development Association *Damp proofing courses and flashings with brickwork & blockwork* Practical Note 6, January 1975.
A brief pamphlet on dpc selection and detailing with some useful comment on dpc characteristics and formation.

4 ● British Standards Institution *Code of practice for walling: part 1: brick & block masonry* CP 121: Part 1: 1973.
The reference for the 'official' word on brickwork and blockwork with brief sections on dpc detailing with some illustrations. See section 4.5 of this book for criticisms of it.

5 ● British Standards Institution *Materials for damp proof courses, metric units* BS 743: 1970.
The reference for the 'official' list of dpc materials but of limited use as it doesn't include some of the most popular dpc materials nor details of performance; under revision: check latest information.

6 British Standards Institution *Protection of buildings against water from the ground* CP 102: 1973. Amendment CP 1511: 1974.
Discusses damp proof membranes and tanking in addition to wall dpcs at ground level. Of limited use as it is mostly in general terms with few illustrations.

7 British Standards Institution *Code of practice for thermal insulation of cavity walls by filling with UF foam* BS 5618: 1978.
Although most of this Code relates to UF foam, the appendices deal with the calculation of the Exposure Index E. The revision of this calculation is under consideration.

8 Building Research Establishment. Current Paper CP 86/74 *Window to wall joints* September 1974.
Some discussion of alternative dpc details at windows but only really covers vertical dpcs at window jambs.

9 Building Research Establishment *Damp proof courses* Digest 77 (second series). December 1966.
This brief pamphlet extends the list of dpc materials to include pitch polymer and in situ coatings with some discussion of material properties. Much of the advice is questionable and only a few well known details are illustrated.

10 Building Research Establishment. Digest 176 (second series) *Failure patterns and implications* April 1975.
Contains a brief mention of dpc detailing and execution as the cause of water penetration to buildings investigated by the advisory service during the period 1970-1974.

11 Building Research Station Advisory Service *Protection from rain* April 1971.
This pamphlet covers a large subject very briefly but is of interest as it spotlights past water penetration problems.

12 Butterworth, B *Brickwork: efflorescence* Structural Clay Products, SCP 5. January 1974. A useful, if slightly eccentric, discussion of efflorescence in brickwork generally. Discusses the basic facts regarding the causes particularly in relation to parapet walls, copings, etc.

13 Department of the Environment, DED Advisory Leaflet 23. *Damp proof courses* December 1966.
A very brief pamphlet on dpcs basically written for the small builder and the lay public.

14 Duell, J *The selection of damp proof courses for load-bearing brickwork—architectural and constructional considerations.* Proceedings of the British Ceramic Society, 24 September 1975.
This book on dpcs is an extension of this paper presented at the fifth international symposium on load-bearing brickwork sponsored by the Building Materials section of the British Ceramic Society, London, November 1974.

15 Foster, D *Brickwork: durability* Structural Clay Products, SCP tn 3. March 1971. A general document on the durability of brickwork but little reference to dpc detailing.

16 ● Foster, D and Allwright, D *Brickwork: resistance to rain* Structural Clay Products. SCP tn 5. July 1973.
This 60-page booklet gives a thorough look at dpc detailing in brickwork with useful tables and details. Although it varies in some respects with recommendations in the present series of articles, it is complementary to it.

17 ● Freeman, Ian *Building failures* DOE Construction 14. June 1975.
This article by the head of the Building Research Advisory Service is related to the report on building defects noted above but is morc positive and gives some suggestions for dpc details including cavity dp trays with corners and junctions similar to those shown in this present book.

18 Greater London Council *GLC Good practice details* Architectural Press 1980.
Illustrates constructional details of walls and roofs, with particular reference to low-rise housing. Little discussion of *why* the details should be done as shown, but some useful 3-D illustrations of dormers, etc. A companion volume *GLC Detailing for Building Construction* has few details relating to dpcs.

19 ● Handisyde, C et al *Everyday details* Architectural Press, London.
The 'Everyday details' series is essential as a starting point for considering dpc detailing as it discusses alternative approaches and highlights possible problems in relation to construction, etc. It shows that even the simplest detail needs careful thought. It is complementary to the 'detailing' section of this present book and particularly relevant cross-references have been given.

20 Knight, T. L. *Illustrated introduction to brickwork design* Brick Development Association. November 1975.
A beautifully produced booklet with coloured diagrams giving a brief introduction to all aspects of brickwork design. An essential reference for students and a useful reminder for others.

21 ● Lacey, R. E. *Driving-rain index* Building Research Establishment Report. HMSO. 1976.
Although published primarily to estimate exposure for cavity fill insulation, the large scale maps and indexes can be applied to other design matters including dpcs. The tables, unfortunately, are only tabulated for buildings 10 m and under in height. This booklet largely supersedes BRS Digest 127.

22 *Specification* 1982 D Martin ed. See 'Bricklayer' section, pp2-94 to 2-118. Useful for its general discussion of brickwork, dpcs, etc and for its lists of proprietary materials and components.

23 National Building Specification *Brick/block walling* Section F 21.
The full version of the NBS is only available by annual subscription. The small jobs version in one volume can be bought separately, and section F 21 is virtually identical in both versions. Although the NBS is a good basis, it is very important that all sections are carefully scrutinised, and amended or extended for the particular project.

24 West, H et al *The performance of walls built of wire-cut bricks with and without perforations* British Ceramic Research Association. Special Publication 60. 1968.
This is a very interesting report on extensive laboratory tests on strength and water penetration in properties of sample walls built in various bricks and mortars.

25 Greater London Council *GLC Preambles to bills of quantities* Architectural Press 1980.
Useful as a comparison with the NBS for specification of dpcs and in general.

Note 1: Refer also to current Agrément Board Certificates (from the Agrément Board, PO Box No 195, Bucknalls Lane, Garston, Watford, Herts WD2 7NG) on various dpc materials, and the GLC *Development and Materials Bulletin* (enquiries to Room N272, County Hall, London, SE1 7PB) for reports on some dpc materials.
Note 2: Especially useful key references are marked thus ●

Index